Eastern Africa Series

PASTORALISM
& POLITICS IN
NORTHERN KENYA
& SOUTHERN
ETHIOPIA

Eastern Africa Series

*forthcoming

Pastoralism & Politics in Northern Kenya & Southern Ethiopia

GÜNTHER SCHLEE
& ABDULLAHI A. SHONGOLO

JC JAMES CURREY

James Currey
www.jamescurrey.com
an imprint of Boydell & Brewer Ltd
PO Box 9, Woodbridge, Suffolk IP12 3DF (GB)

and of

Boydell & Brewer Inc.
668 Mt Hope Avenue, Rochester, NY 14620–2731 (US)
www.boydellandbrewer.com

The publisher has no responsibility for the continued existence or accuracy of URLs for
external or third-party internet websites referred to in this book, and does not guarantee
that any content on such websites is, or will remain, accurate or appropriate.

A CIP catalogue record for this book is available from the British Library

ISBN 978-1-84701-129-9 (James Currey Paperback)

This publication is printed on acid-free paper

Typeset in Cordale by CPI Typesetting

Contents

v

List of Maps and Figures

List of Abbreviations

AU	African Union
B.	Boran (Language)
BORE	Boran/Rendille
CDF	Constituency Development Fund
CIFA	Community Initiatives Facilitation Assistance
COPEP	Community Peace Programme
DC	District Commissioner
EPRDF	Ethiopian People's Revolutionary Democratic Front
FORD	Forum for the Restoration of Democracy
GGA	Gumi Gaayo Assembly
IDP	Internally Displaced Person(s)
KANU	Kenya African National Union
Ksh	Kenya Shilling
LASDAP	Local Authority Service Delivery Assistance Programme
LATF	Local Authority Trust Fund
MP	Member of Parliament
NARC	National Rainbow Coalition
OALF	Oromo Abo Liberation Front
OLF	Oromo Liberation Front
OCPD	Officer Commanding Police Division
OPDO	Oromo People's Democratic Organization
PARIMA	Pastoralist Risk Management
PISP	Pastoralists Integrated Support Programme
PC	Provincial Commissioner
PRS	Proto-Rendille-Somali
R.	Rendille (Language)
REGABU	Rendille, Gabra, Burji (suspected ethnic coalition)
SACCO	Savings and Credit Cooperative Organisations
SALF	Somali Abo Liberation Front
Som.	Somali (Language)
Sw.	Swahili (Language)
TOLs	Temporary Occupation Licences
TPLF	Tigrayan People's Liberation Front

Introduction
GÜNTHER SCHLEE

There is no need to worry about the future of pastoralism. Pastoralism came into existence thousands of years ago, not long after mixed agriculture from which it has derived as a form of specialization. It can be temporarily obstructed, oppressed, or abolished but it will always re-emerge. The reason for this is simple: about one third of the land surface of the world and two thirds of Africa (United Nations 1997) are arid or semi-arid and cannot be used for any other form of food production.[1] Low rainfall and its erratic distribution will always require herd mobility. If the groups utilizing these areas now, the so-called 'traditional' pastoralists are exposed to political and economic conditions which reduce them to poverty and force them into sedentarization or if they are destroyed by military force, others will take over. 'Modern' pastoralists of urban or agricultural origins might play that role, assisted by satellites (remote sensing, GPS) and other modern communication technologies. The existing groups of pastoralists can be expropriated, marginalized, expelled, or decimated. But then non-pastoralists will become, as they have done again and again throughout history, the new pastoralists. The question is not whether also in the future there will be a mobile form of livestock production. There will. The question is to whom it will belong.

In the case of Kenya, huge areas of former pastoral land belonging to the Maasai were cleared of Maasai and given to white ranchers. In the First World War, the British and Africans from Kenya, then the British East Africa Protectorate, had been fighting side by side against the Germans and 'their' Africans in Tanganyika, but when demobilization came and soldiers had to be settled, land belonging to the Africans was given to the Whites. Pastoral land became ranch land. After independence some ranches were sold to the new Kenyan elites, who – unsurprisingly – were of almost exclusively sedentary background, not of pastoralist background, and stemmed from more developed, more densely populated parts of the country with better educational facilities and more economic and political clout. But the new ranchers soon discovered that their ranches

---

[1] Food production in this sense refers to the activities which date back to the Neolithic revolution and the domestication of plants and animals. Hunting, gathering and fishing, according to this definition, are not food production but food acquisition, since the organisms put to human use in these modes of livelihood have not been assisted in their proliferation and growth by the efforts of humans.

were too small to balance the risks involved in rain falling in one place and not in another. Universally, the amount of rainfall is correlated to patterns of distribution. The lower the precipitation the more erratic it is. There is no reliable minimum. The ranchers soon started to lease pastures to each other and to trek or truck cows across the country. Livestock production resumed its nomadic nature. But cows now belonged to the ranchers, not to the Maasai. Settler families sent their children, typically two or so, to expensive boarding schools in England. The land on which hundreds of low-cost Maasai children were raised now served to earn the fees for two children and these fees were payable abroad. So again, the question is not whether livestock production is mobile or not but to whom it belongs and where its proceeds ultimately go.

In Kenya it is largely the northern half of the country plus much of the hinterland of the coast in the east and the Masai steppe in the south which is arid or semi-arid and used by pastoralists. Also, the Ethiopian highlands are surrounded by pastoralist lowlands on all sides. The present volume is concerned with those in the south which geographically and ethnically form one zone with northern Kenya.

Although this vast lowland covers the larger part of the surface of Kenya, in the minds of many people it does not really form part of it. When weary travellers, shaken up on a lorry from Moyale[2] or Marsabit after hundreds of miles of rough, corrugated dirt road reach the tarmac at Isiolo and the shaking suddenly stops,[3] they say, with an element of irony: 'Now we are in Kenya!' Also the central Kenyans do not really perceive of the north as a part of their country and often do not have the faintest idea about its nature. When once, hitting town thirty years ago, I tried to explain to a group of ladies in a bar in Nairobi that I did field research among camel herders in northern Kenya, they wanted to make sure that I was not talking about Arabia. They were surprised to hear that there are camels in Kenya. The problem is not with these ladies. The problem is that the political class in Kenya often does not have a much better appreciation of the character of these remote areas of their country and the issues at stake here.

Over the decades I have had a huge number of similar experiences, but I have not systematically documented them. To some extent the press-cuttings collected by Abdullahi Shongolo, a selection of which we quote in this volume, mostly in the chapter on 'Moi Era Politics', make up for this. In many of these one can find stereotypes, misrepresentations and expressions of ignorance in the minds of central Kenyans about the pastoralists in the north of their country. Lacking a detailed documentation of misrepresentations of pastoralists by non-pastoralists from my own experience, let me cover this aspect of the Kenyan social reality by pointing to the work of Saverio Krätli (2006).

[2] Following some maps, here we use Moyale for the town as a whole or, more specifically, for the Kenyan part of the town, and Moiale for the Ethiopian part

[3] For the last couple of years the tarmac has been pushed northwards and now, 2011, it reaches the southern end of what used to be Marsabit District.

Krätli (2006 pp. 123–8) addresses the question whether the poverty of pastoralists – in this case he studies the Turkana and Karimojong on both sides on the border between Kenya and Uganda – can be found to have cultural roots. The answer is 'yes' but with a new twist. While common wisdom attributes poverty to the 'traditional' or backward 'culture' of the impoverished marginal groups themselves, Krätli finds the problem in the 'culture' of the mainstream. There is a culture of misrepresentation of pastoralists in the popular, political, and 'development' discourses of the dominant society which has detrimental, marginalizing and impoverishing effects on those talked about.

Mainstream culture in school books, press reports, policy statements and NGO discourses depict pastoralists as warlike, primitive, and backward – if they are discussed at all. Taxi drivers in Kampala appear to have as vague an idea about where Karamoja is located as the Nairobi ladies cited above have about where to find camels.

A particularly illustrative example about a certain normative image of modernity held against pastoral mobility is what Krätli reports about corrugated iron sheets. Corrugated iron sheets for the roofing – and sometimes also for the walls – of immobile homes seem to be the epitome of modernity which is as closely linked to a sedentary form of life as nomadism is – wrongly – identified with an early stage of development. Krätli describes how some tarpaulins from lorries from a closed down World Food Programme outpost had found their way into Turkanaland. The pastoralists found them very practical for the construction of temporary shelters and wanted to buy more tarpaulins. This idea was ridiculed by a Turkana official who said that his tribesmen should procure themselves corrugated iron sheets rather than tarpaulins (Krätli 2006 p. 131).

Contrary to the image of being conservative and hostile to innovation, Krätli found pastoralists very receptive for things which they find practical, like rain coats. I can confirm this, as many other observers can. Mobile phones in the early 2000s made rapid progress in pastoral populations in Africa. The possession of phones spread faster than the networks. Where the networks on the ground were too weak, people climbed trees[4] or hills or went to specific locations where, for unexplained reasons, the telephone connections were better than in the immediate neighbourhood.[5] There must now be thousands of places across Africa which are called 'Network'.[6] Where the network was weak close to the ground, people ingeniously hung phones from the ceilings of their huts, put the loudspeaker of the phone on in order to hear the other participant and stood on their toes to answer.[7] There was certainly no conservatism at work which slowed down the reception of this particular

---

[4] Oral communication by Elhadi Ibrahim Osman about Mbororo Fulbe along the Blue Nile in the Sudan.

[5] Own observation from Sennar State, Sudan.

[6] Own observation from Kenya and the Sudan.

[7] Own observation, Sudan.

Figure 0.1 Peak Revision, KCPE Science book. This Hereford cow from the cover of a Kenyan Primary school book may send the pupils, including the children of pastoralists, the signal that the ways African cows are managed on African dry lands are not worthy to be studied or are not part of the knowledge needed in the modern world. Education which does not relate to the pastoralist reality alienates the children of pastoralists from the pastoralist way of life and way of production, not just because it is provided in a sedentary form rather than being brought to the pastoralists, but also because of its content. *(Source: Krätli and Dyer 2009 p.53)*

modern technology.[8] We can only agree with Krätli's finding that pastoralists are fast in adopting innovations which are useful to them and that their reluctance to adopt other innovations might have to do with their lack of usefulness.

In contrast to the backward image projected on them, pastoralists do not perform so badly along a number of parameters. Citing Scoones (1995 Table 6.1), Krätli points out 'in Kenya and Uganda returns per hectare and per animal are higher in "traditional" systems than in ranching' (2006 p. 127). Farmers in industrial countries receive huge subsidies, even in spite of the disadvantages to which they are exposed (like 'quarantines' favouring the ranchers whenever pastoral competition on the market is too strong (Schlee 1990b)). This juxtaposition leads to serious questions. Who is better? Which system of production is more efficient? In a similar context, Krätli points out that the subsidies to farmers in OECD countries exceed the GDP of Africa (Krätli 2006 p. 134). Subsidized

---

[8] Hussein A. Mahmoud (2003) stresses the importance of mobile phones for livestock traders. Cf. the chapter on 'Ecology and Politics' in Schlee with Shongolo (2012).

farming leads to overproduction in OECD countries. Excess grain then is sometimes sent as food relief to Africa.[9] This food relief is perceived as a humanitarian intervention and an indicator of economic failure. The subsidies, however, which have led to global inequality and distorted markets, are neither seen as humanitarian intervention nor as indicators of failure. Why not?

Krätli gives many examples about stereotypes which misrepresent pastoralism and lead to their marginalization and to misconceived development intervention from his experience in Kenya and Uganda. He could well have adduced more examples from the same book (Dyer 2006) in which also his own contribution, cited above, is included. The chapter by Carr-Hill (2006) abounds with unverifiable stereotypes which have wide currency in development circles.

In his chapter, which is about 'Educational Services and Nomadic Groups' in six East and North-east African countries, Carr-Hill seems to regard the poverty of pastoralists as an unquestionable given. After defining nomads he moves on straight to defining poverty, a concept which does not even appear in the title of the chapter. Not a single line is wasted on explaining why a discussion of nomadism should lead to a discussion of poverty. Apparently, this happens automatically.

The data given is then used to make the point that pastoral economies have been marginalized and have become dependent on remittances from family members in town. Still, there is a bias in reporting pastoral poverty, making it look worse than it is. If it is such an unrewarding way of life, why do even impoverished and sedentarized nomads proudly maintain a pastoral social identity and put a high value on having herds, as Carr-Hill also reports (2006 p. 39). Monetary income (GDP per capita) is compared and found to be only half as high for the pastoralists as in the national average,[10] and subsistence agriculture is briefly mentioned as an economic factor in the lives of agro-pastoralists (Carr-Hill 2006 p. 41). But the obvious fact that herd owners derive subsistence from their herds, that they drink milk and slaughter their own animals, at least on ritual occasions, is not mentioned at all, apparently because this subsistence use of animals is normally not expressed in monetary terms. A similar 'modernist' perspective informs the statement that 'lack of high-grade stock' (Carr-Hill 2006 p. 40) is one of the problems pastoralists have. High performance breeds from Europe have been extensively used in Africa and, by cross-breeding, their genes have spread widely in higher and lower proportions. Breeding for special performance features, whether on the basis of European breeds (called 'exotic' in Africa) or of local stock, however comes at a price. Fast growth comes with higher demands and lower resistance to deprivations. Nomads in arid environments (and even

---

[9] The 'good practice' of food relief would be to buy grain from areas as close as possible to the areas in need of it so as to minimize the adverse effects of free distribution of food on food production in a region. This practice is not always followed.

[10] 100 USD as against 200 USD for Eritrea (Carr-Hill 2006 p. 40).

villagers who use unimproved pastures around villages) therefore often have preferred to keep their local unimproved breeds or have reverted to them after acquiring experience with 'improved' stock.[11]

DECONSTRUCTING THE ARCHAIC PASTORALIST

Popular perceptions of pastoralism, including perceptions by politicians and development planners, too often seem to be informed by theories which date back at least to the eighteenth century. The classical evolutionist three-stage-model of the type hunters-herders-farmers had nomadic livestock production developing straight out of hunter, or as we would now say hunter-gatherer, forms of livelihood. Mobile herders were believed to precede sedentary agriculturalists with a mixed economy and were thus more 'primitive' than these latter. E. Hahn (1891, 1892, 1896, 1911, 1913, 1925, 1927) may have been the first to doubt this sequence. He pointed to something rather obvious, namely that it is difficult to domesticate herbivores in a hunting economy. To feed captured young animals, at least some early form of agriculture is required. The only apparent counter-example, discussed in the older literature, the domestication of reindeer by taiga hunters, may not have happened independently but may have been derived by stimulus diffusion from the steppe where cattle and horses had been domesticated for a long time (Vajda 1968). For quite some time now, the consensus that the domestication of ungulates has never and nowhere taken place independently or prior to the domestication of plants has been very broad. Directly or indirectly all forms of livestock keeping have been found to derive from mixed farming. Rather than being 'primitive', mobile pastoralism is a comparatively recent and rather sophisticated specialization out of a mixed economy.[12]

There is a more recent debate which seems to rehabilitate Montesquieu and the idea that nomadic pastoralism is a very old form of production. This debate relates to the question of independent domestication of cattle in the Sahara and whether or not it preceded the domestication of cattle elsewhere and – more importantly – whether or not it preceded the domestication of plants and crop production. If the latter is confirmed, the eighteenth century theorists who portrayed 'nomadic' pastoralism as directly evolving from 'nomadic' hunter-gatherers would, after all, be right.

This is how Homewood (2008) summarizes the new finding about the eastern Sahara:

> From around 12000 BP, perhaps driven by changing climatic and environmental conditions, people in north-east Africa, south-west Asia and the southern part of East Asia all independently invented the domestication

[11] Oral information about Kenaana cattle in the Blue Nile area of the Sudan by Awad Karim Tijani. Cf. also Schlee (1988b) on Rendille and Somali camels.

[12] For a fuller discussion see Schlee 2005.

of animals and plants for food. In each case, the process seems first to have centred around the domestication of a focal animal species, with local pre-agricultural traditions of wild plant management later refined into production of domesticated plants. Over the next few thousand years another 7–10 independent centres of domestication arose world-wide, including several more in Africa. The present discussion focuses on the origins of the livestock species which underpin present-day African pastoralist societies.

African ecoclimatic zones and conditions of 12000–11000 BP resembled those of the present day. Between 11000 BP and 9000 BP climates were warm and wet, Lake Chad reached its maximum extent, and perennial watercourses flowed from the Central Saharan Tibesti and Hoggar massifs. During this period, Khoisan-speaker traditions were found throughout the eastern and south-eastern wooded savannas. The Afrasans were associated with Mediterranean climate regions throughout Northern Africa and the Red Sea Hills. Sudanic-speaking Nilo-Saharans spread north as savanna vegetation expanded northward into the Sahara. Coming into interaction with the Afrasan peoples, they adopted the use of wild grains, and transposed the techniques of wild sedge and grass collection and preparation to new species of wild cereals including fonio, and wild forms of sorghum and pearl millet. The moist warm period also meant a southward extension of Mediterranean climates, habitats and faunas including Bos primigenius africanus, the wild ancestor of African (and European) indigenous cattle domesticates, and present in North Africa into historical times. The Saharo-Sahelians of the Middle Nile are thought to have domesticated cattle first. They left domesticated cattle remains dating to 9400–9200 BP, alongside evidence of North Sudanic wild grain use (Nabta Playa – Wendorf et al. 1984, 1987; Wendorf and Schild, 1998, 2001; and Bir Kiseiba: Egypt – Gautier 1986, 1987; Marshall, 1994; MacDonald, 2000; Blench and MacDonald, 2000). These pre-date the first food-producing economies of the more northerly Nile Valley and Delta (Holmes, 1993; Stanley and Warne, 1993), and the first domestic cattle in south-west Asia, by a thousand years or more (cf. Russell et al. 2005). Nilo-Saharan invention of ceramics (before any Middle Eastern or European pottery) underpinned their development of cooked porridges and gruels (in contrast to the Afrasan baked breads).

North Sudanian culture and demography were gradually transformed by their cattle keeping. Between 10000BP and 9000BP these Sudanic people began cultivating sorghum and millet derived from their wild grains, and later gourds and cotton as well, developing spinning and weaving. (Homewood 2008 pp. 14f)

This account is debatable on several levels. We start with Homewood's central point, which is also the one which has the most far-reaching implications for what we have said so far about pastoralism being a form of specialization out of mixed agriculture which included plant cultivation. Homewood diametrically contradicts this position by claiming chronological anteriority of the domestication of livestock. She does this on a world-wide scale, including the Eurasian centres of domestication. In each case the domestication of a focal species of animals is said to precede the domestication of plants. She gives no references whatsoever to underpin this far reaching claim. She focuses on the African case. As there were no other sources to check, I therefore checked the sources she cites for the domestication of cattle in north-east Africa.

Her sources do not say what she claims they say. Rather than claiming that the early pastoralists (whom Wendorf and Schild perceive in their remains in the Egyptian Sahara) predate agriculture, they state: 'Preliminary chemical analyses by infrared spectroscopy of the lepids [sic; lipids?] in the archaeological sorghum show closer resemblance to some modern domestic sorghum than to wild varieties' (Wasylikowa et al. 1993). Along this same line it is interesting to note that the distribution of the sorghum in the houses suggests that sorghum was treated differently from the other seeds. The significance, however, is not in whether or not the sorghum was wild or domestic, but that the sorghum and other plants were being intensively harvested and stored for future use. One may conclude that plant foods comprised a significant portion of the El Nabta diet' (Wendorf and Schild 1998 p. 104).

In a later publication (Wendorf, Schild et al. 2001 p. 8) the statement that 'sorghum was treated differently from the other grasses and may have been cultivated, although it was morphologically wild' is reiterated.[13] In fact, the point Wendorf and his colleagues always wanted to make is that both plant domestication and the domestication of cattle are old in the Western Sahara and independent of other centres of domestication, apart from the nearby Nile Valley, not that one predates the other (Wendorf et al. 1992).

In fact, they do not even locate the domestication of cattle in these Saharan site at all but 'suppose that these cattle were first domesticated at an earlier, but unguessable, date in the Nile Valley' (Wendorf et al. 1987 p. 447). In the sources I could check I have not found a single line indicating whether the people who first domesticated cattle ate wild or domestic grain or both.

What is clear, however, is that the Saharan cattle keepers had lots of grains and tubers and legumes and had surpluses to store beyond their immediate needs. If the so far inconclusive debate about whether the grains included products of cultivation should one day come to the conclusion that all these plants were wild, and that also the original domesticators of cattle in the Nile Valley were not cultivators but hunter gatherers, our earlier statement that pastoralism is a specialization out of earlier mixed agriculture would indeed have to be qualified by an exception. On the other hand the collection of wild grass seeds (grain) among all forms of gathering is the one which comes closest to agriculture. In the Sahel even today wild grain is collected by swinging baskets across the stand of the tall grasses. Some grains fall into the baskets while others bounce off and are spread all around. This is a symbiosis between humans and grasses, both helping each other to proliferate, which comes very close to agriculture.

Our point against the early evolutionists, who claim that nomadic

---

[13] The argument about domestic traits of the plant remnants seems to be debatable. Cf. Mathilda's Anthropology Blog (2009).

pastoralism evolved directly out of an economy of hunter-gatherers, was that animals cannot be domesticated by hunting them. One needs surpluses from agriculture to raise young animals one has caught. We cannot decide whether a pure gathering type of economy can generate enough surpluses beyond the needs for human consumption to make this possible. But we do know that the people of the early Holocene period in the Western Sahara did have enough grain to store it, irrespective of whether they grew or just collected it.[14]

When Homewood characterizes *Bos primigenius africanus* as the 'the wild ancestor of African (and European) indigenous cattle domesticates', she cannot possibly mean that European cattle derive from *Bos primigenius africanus*. From *Bos primigenius*, the aurochs, yes, but not specifically from *Bos primigenius africanus*. Some specifically African mitochondrial genes have been found in Iberian cattle, but these have not shaken Ajmone-Marsan's and his colleagues' assumption that European cattle derive from cattle domesticated in Anatolia which has subsequently interbred with wild *Bos primigenius* within Europe, and this seems to be a widely held view among archaeologists. The African genes found in Iberian cattle 'can be traced back to either the Moorish occupation or prehistoric contacts across the Strait of Gibraltar' (Ajmone-Marsan et al. 2010 p. 150, giving four sources). There seems to be no evidence that European cattle in general derive from stock domesticated in Africa.

On the whole, it must be said that, irrespective of the still open debate about independent domestication of cattle in Africa and its anteriority to other centres of domestication (as an Africanist I sympathize with the idea), there seems to be nothing to support Homewood's statement that cattle domestication in Africa and elsewhere preceded plant domestication, i.e. agriculture. We can therefore, until compelled by evidence to the contrary, maintain our position that nomadism and other forms of mobile pastoralism are a specialization out of a more diverse economy, namely mixed agriculture. Mobile pastoralism is a highly sophisticated and strategic form of use of often rather extreme habitats. There is nothing archaic about it. Specializations always develop out of more generalist ancestral forms, and mobile pastoralism is just one more of these specializations, and a rather elaborate one. I think basically that is also the point Homewood wants to make, although the flow of her writing and its overflow sometimes lead her astray from that perspective.

---

[14] There seems to be a way to raise young large herbivores without any plant fodder. Rendille told me (I have never actually witnessed it) that in case the mother of a young camel calf dies, her meat can be cut in thin stripes and dried, so that it lasts for months. A soup of it can then be fed to the calf: if there is no mother's milk, mother's soup will do. The amount of meat seems to be roughly nutritionally equivalent to the amount of milk over a lactation period. The Rendille have large earthen cooking pots in which they cook soup. If the first domestication of cattle, assumed by the authors we discuss here to have happened in the Nile Valley, was carried out in a similar way (killing a wild cow and raising the calf with her meat), this presupposes pottery. Especially for the early parts of their archaeological record Wendorf and Schild (1998 p. 100), however, speak of pottery being rare and a luxury item.

Although the finding that pastoralism is not 'archaic' but a fairly recent development (on a scale of thousands of years) has become quite widely accepted in the academic world, it is surprising how tenaciously the image of 'archaic' pastoralists is held in popular opinion. Leder and Streck (2005 p. xi) would describe this as being part of the 'ages-old history of stereotypes produced to express sedentary people's mistrust of nomads.' Such culturist perceptions of nomadism owe some of their tenacity to the fact that they have been combined with 'scientific' positions, which, of course, are also part of 'culture' in a wider sense.

When the British introduced tribal grazing areas and quarantine belts in colonial Kenya they took these measures which limited pastoral mobility for reasons which they believed to be scientific. They were convinced that the land was 'overgrazed' and therefore exposed to 'erosion'. Apart from restricting all groups to 'their own' areas, so that the consequences of over-utilization would be suffered by those who caused it, 'de-stocking' was thought to be the remedy. Chenevix Trench (1993) describes the disastrous 'sheet erosion' caused by 'overgrazing' and the beneficial effects of colonial policies in Samburu. But already Spencer's (1973 pp. 180ff) account shows a much more complex picture and can by no means be read as a success story.

Rapid environmental decline seems, however, to have been a dogma among expatriate scholars as well as African bureaucrats and has justified numerous studies and all sorts of interventions. For someone like me, who has visited the same areas in northern Kenya for over thirty years, the empirical base for these assumptions seems to be lacking. Most areas look roughly the same as ever, and localized destruction of the vegetation through overuse or trampling seems to be due to lack of mobility and the concentration of stock around major settlements rather than the open character of the range and 'the tragedy of the commons'.

Apart from such localized forms of vegetation destruction that impairs re-growth, the regeneration of pasture in northern Kenya seems to depend exclusively on the highly erratic rainfall. While the British thought that there was an equilibrium between herbivores and pasture, which needs to be maintained by limiting grazing pressure to the 'carrying capacity' to guarantee the re-growth of the same plant biomass next year, range ecologists have since then found that in settings like semi-arid northern Kenya the growth of plants in one year has little to do with how many animals fed on them in the preceding year. Instead it is proportional to the amount of rainfall. Environments in which this is the case have been called 'disequilibrium environments', maybe somewhat unfortunately because it really is about the absence of an equilibrium not the presence of a disequilibrium (Behnke, Scoones and Kerven 1993).

Another 'scientific' reason for restricting nomadic movements was livestock epidemics. All across Kenya there was a quarantine belt separating the nomadic north from the 'developed' part, and even today it frequently occurs that livestock transports from one District to another are forbidden. In many cases, however, the same diseases were endemic on both sides of the barrier, making the scientific reason of quarantines questionable. Looking for an explanation, one should not forget the broader picture. Pastoralists are not the only producers of livestock in Kenya. There was, and to some extent is, a White Settler economy based on large ranches (Raikes 1981, Schlee 1990b, p. 1998a).

It is obvious that the quarantine regulations shielded the ranch sector against competition from pastoralists to the benefit of white settlers and later, after independence, of the new elites who took over many of the settlers' ranches. The quarantines were part of a dual livestock economy with a low price and a high price sector. Livestock was accorded different grades, and African-owned livestock was accorded the poor grades and had to be purchased by ALMO, the African Livestock Marketing Organization, the colonial precursor of the KMC, the Kenyan Meat Commission. The 'auctions' carried out by these organizations did not deserve that name, because prices per kilogramme of live weight were fixed at a low level.

The colonial literature is full of laments about the irrationality of the pastoralists, who did not want to sell and kept unproductive animals for sentimental reasons or for prestige. The facts are that no group of pastoralists in Kenya has ever been self-sufficient. They always engaged in barter for agricultural produce, they had glass beads from as far as Venice or Bohemia and cowry shells from the Indian Ocean. Whether or not they sold animals always depended on the price, and not to sell for an artificially fixed low price should have been recognized as quite rational also by these colonial writers, and as corresponding to their own market behaviour in a similar situation.[15]

Still, although obviously defying an economic logic, the territorial regulations imposed by the colonial government were phrased in economic terms and in terms of 'development'. Galaty (2009) cites a number of reports and plans from the 1950s, which linked land ownership to agrarian development and justified the implementation of land registration in the Kikuyu reserves in the same way as the demarcation of sectional boundaries and later group ranches among the Maasai.[16]

'Scientific' ideas of pastoralism as an ecologically harmful and economically inefficient practice have by now become a matter of the past. For quite some time views that depict pastoralism as a sophisticated and strategic, as well as sustainable, form of using resources which are difficult to use otherwise have also found their way into official documents, like the Range Management Handbook of Kenya (Shaabani et al. 1991). A more recent example would be a publication by the International Institute for Environment and Development and SOS Sahel (IIED and SOS Sahel 2009) which specifically addressed policy makers. Actual effects of such writings on policies have, however, so far been found wanting.

[15] For examples of stereotypes about pastoralists see Schlee (1989b pp. 398 ff).

[16] The preceding paragraphs have been adapted from Schlee (2010).

A measure of hope for change in actual policies can be attached to a new document issued by the African Union (AU), namely 'A Policy Framework for Pastoralism in Africa' (adopted by the AU January 2011). It proposes to develop pastoralism rather than defining development as pastoralists quitting pastoralism to become something else. It gives this agenda a continental scale by pointing out that 40 per cent the African land mass is used by pastoralists and that these 'pastoralists help to protect and safeguard the resources found in these arid and semi-arid areas' (African Union 2010 section 1.1.1).

A key issue the AU Framework addresses in the context of pastoral mobility is land rights. It states that the collective land rights of pastoralists need to be protected against infringement by claims to private property and alternative forms of land use (African Union 2010 section 1.1.2). These alternative forms of land use comprise sedentary crop production. As crop production in rangelands tends to be possible only in certain favourable spots which also provide the richest and most reliable pasture, it is these most productive parts which are taken out of the pastoralist cycle of migration. This reduces the productivity of the remaining pastoralist economy greatly. These choice parts of the pastoralist range which are 'allocated to private companies for commercial agriculture' frequently lie 'in riverine areas which are often critical dry season resources for pastoralists'. The document also mentions large scale irrigation schemes in this context (AU 2010 section 3.2.1). Although no specific examples are given, it is difficult not to think of Ethiopia at this point. There the government has declared large chunks of pastoral land to be unoccupied or simply expropriates the lands of pastoralists and agro-pastoralists and leases it to foreign investors. While observers are focusing on the Gibe III dam project on the upper Omo which is going to destroy local economies further down which are based on floods and flood-retreat for cultivation, the alienation of land belonging to agro-pastoralists and pastoralists for large-scale irrigated agriculture is also taking place along the Blue Nile and the Sobat (Gambela Region) and wherever there are even surfaces close to bodies of water.

The document also recognizes that pastoralists in a strategic and knowledge-based way combine the use of resources on different sides of boundaries, including national boundaries, and that this cross-border mobility needs to be preserved and protected. Personally, I find great relief in reading this. My experience from northern Kenya is that cross-border movements of livestock are primarily seen by administrators as leading to the spread of diseases even when it is plain that the same diseases are endemic on both sides of the border (see above on 'scientific' reasoning and quarantines). With the elite-driven tendency to territorial subdivisions to form new and smaller ethnic-based Districts (which will be one of the main topics dealt with in subsequent chapters) even District boundaries have turned more and more into obstacles to pastoralist movements. The AU document may be well-timed to fight these tendencies.

The framework states the dual objective of improving the lot of the pastoralists by protecting their 'lives, livelihoods and rights' and of advancing the 'national, regional and continent-wide economies' by reinforcing the 'contribution of pastoral livestock' to these economies (AU 2010 section 1.2). Already now, pastoralists 'in many African countries supply the bulk of livestock for domestic meat markets' and there is a well-established and robust livestock export trade that continues to grow and respond to new market opportunities (AU 2010 section 2.1). Although this may not be new, it is an important statement, because many urban Africans, including the political class, do not seem to be aware that the relatively cheap and abundant, as well as 'ecologically' produced,[17] meat they enjoy does not stem from the 'modern' kinds of agriculture they propagate but from pastoralists. Further down (AU 2010 section 3.2) the document makes the comparison of the relative merits of 'modern' agriculture and pastoralism quite explicit: 'A major part of the move towards modernization was the commercialization of agriculture, characterized by the shift from subsistence to commercial farming, from pastoral communal ownership to privatization of pastoral land, and from pastoral traditional institutions to seemingly modern ones. Within these frameworks pastoralists' mobility continued to be viewed as irrational and economically inefficient, despite the emerging body of research which demonstrated that pastoralist production was at least as efficient as modern ranching.' Finally, pastoralists have found their way into official perception as effective producers. Too long has a discourse been dominant which represents them only as recipients of humanitarian and emergency aid, a discourse which does not take into account that, if pastoralists had had a fair deal in the first place, they might, in spite of the unpredictability of their environment, not need humanitarian aid nearly as often.

The document does not fail to notice some positive trends. The Pastoral Code in Mauritania and the Pastoral Charter in Mali recognize and protect pastoral mobility. The Government of Chad has introduced a mobile school system, while the Kenyan Ministry of Education is pursuing a programme of distance learning. Since 2008, Kenya has a Ministry of State for the Development of Northern Kenya and other Arid Lands. Already in 1998, ECOWAS set up a regional framework for cross-border transhumance between fifteen member states. There is enough good news of this kind to fill a page and a half (AU section 3.2.2).

To preserve pastoral mobility, incentives to sedentarization need to be viewed critically. Schooling and other services, when provided on a sedentary basis, are such incentives of sedentarization (besides alienating pupils from their reality as producers and preventing them from learning the complex skills of the business). Often, pastoralist families face the dilemma of having to choose between schooling for their children – or health-care for an ailing family member – and effective management

---

[17] There is, however, room for improvement in veterinary services and the instruction of pastoralists in the proper use of antibiotics.

Figure 0.2 Gabra girl loading camel near Farole on the Ethiopian/Kenyan border. Nomadism is a form of specialization which has nothing archaic about it. On the contrary, it is not only sophisticated in the field of organization, but also technologically sophisticated. Animal transport is one of the requirements of mobility. Animals have been trained and the load needs to be fixed in a particular way so that it holds and nothing is left behind. Here a Gabra girl climbs up the load, inside the semi spherical frame made up of the bent sticks used for the mobile houses, to tighten a rope while her mother looks on. *(Photo © G. Schlee 1986)*

of the herd. The document takes up this issue and recommends mobile forms of service delivery. 'In education these include distance learning and alternative basic education approaches; in health, community case management and community health worker systems have been proven to be effective; for basic veterinary care, community-based health workers can be used. In some countries, these and other approaches – tailored specifically to the pastoral context – require policy support if they are to be scaled-up and properly regulated and monitored' (African Union 2010 Strategy section 1.7).

Laudable and timely as this document may be, some problems remain. The first and most obvious one is that there is no direct transmission-belt

Figure 0.3 Loaded camels. The finished load creates a shaded surrounding and thereby provides a space in which babies, elderly people, or lambs can be transported. Apart from two poles on both sides of the shoulders, all other items are part of the house or household items. Rather than having loading saddles or heavy containers, the Gabra tie the load in a way that one part of the load holds and shelters the other. *(Photo © G. Schlee, 1986)*

between those who have formulated this policy and those who are meant to implement it. The authors (the Department of Rural Economy and Agriculture of the African Union) are not part of the chain of command of any ministry in any African state or of livestock development officers, pastoralist elders and NGOs who work further down at District level. It can only be hoped that stakeholders at all levels derive inspiration from this document. Otherwise there is no guarantee that it will have any effect at all.

Being a document which stems from an international body which shares universal values, some contradictions between these values and the realities of pastoralists are bound to turn up. The document identifies the 'need to address age-old rigidities in traditional beliefs and structures, which discriminate against women'. These include equal rights to inheritance (African Union 2010 Strategy section 1.5). Systems of inheritance and allocation (to different wives during the lifetime of the husband) and to the children of these wives are a complex issue. The rights of the husband who is mostly the nominal owner representing the management unit to the outside is limited by particular rights of members of his family or families. Generalizations are difficult and detailed descriptions of systems of ownership would not fit into the space granted to this brief comment. Suffice it to say that before abolishing old systems of ownership and allocation along with the advantages they have to women, one should have a good look at them. Different people have different kinds of rights in

15

Figure 0.4 Camel loading. A Gabra loading a camel in front of a house. The
construction principle of a Gabra house is that of a tent. They consist of a framework
of sticks which have been dried in a bent position. This framework is covered with mats.
(Photo © G. Schlee 1986)

and claims on animals at the same time. Rights of inheritance have to be
seen as systemically interconnected with claims of maintenance and allocations for usufruct. This systemic whole, composed of different rights,
needs to be taken into account when one propagates change. Otherwise
one might end up with a vacuum, in which old forms of security and
entitlement for women have ceased to exist and new forms have not come
into being with the desired effects.

Many African pastoralists are Muslims. According to the *shari'a*
(which, however, is hardly implemented to the letter anywhere), women
should not inherit equal shares but half the share of their brothers. On the
other hand, they do not have to spend their livestock on their own maintenance or that of their children, because (ideally) there is always a man
(a husband, or in the absence of a husband a father or brother) responsible
for their maintenance. So (again theoretically) they can save or invest

Figure 0.5 Stowing a Garre child. A Garre child in Mandera District has been stowed away on a loading camel in a cavity formed by a hide which, when the house is built again, will be used for bedding. *(Photo © G. Schlee 1990)*

whatever they own and on the whole should be much wealthier than men whose resources are constantly diminished by their obligations. In practice, rural women in Muslim societies often do not inherit anything at all, neither land nor livestock. But to improve their lot, appeals to the *shari'a* (which is based on legal differences between the genders and accords duties, rights and privileges in a gender-specific way) in such contexts might be more helpful than trying to introduce an alien concept of equal rights which will not be accepted.

In these domains, it is to be hoped that the document will inspire discussions and detailed examinations of different systems of rights in animals, rather than fast and ill-adapted solutions. On the whole, however, this document, as it stands, can become a powerful tool for the defence of the interests of pastoralists. One can only wish for its wide circulation and that it ends up in the hands of people who make good use of it, as the institutional channels for its implementation appear to be weak.

After introducing pastoralism and some of the problems of how it is perceived in its political context, I now turn more specifically to the present volume and its ethnographic base.

To a much higher extent than our book about *Islam and Ethnicity* (Schlee with Shongolo 2012) which, with a new focus, takes up texts which have been written in the 1980s and 1990s, the present volume is based on research which has been carried out in the present millennium. Some old things, however, remain true and deserve to be taken up again. Articles which I have published in languages like French and German may 17

Figure 0.6 Qur'an class, Wajir District, Kenya. Qur'an teachers among the Degodia Somali move with the nomads. Here the teacher, with the Qur'an in his hands, stands in front of his students. In the background is the shade hastily constructed of branches, under which the class is held. *(Photo © G. Schlee 1989)*

have been beyond the horizon of a broader English-speaking readership anyhow, and there is no indication that this is going to change as the ideal of polyglot scholarship seems to continue to lose ground. I therefore take the liberty of revisiting some old things. Some paragraphs of what I explain about pastoral egalitarianism further down in this introduction can also be found in a French article from 1984 (Schlee 1984b). There is no reason to avoid this repetition, as my findings still seem to be valid. Moreover, my British and American friends and colleagues, whose references tend to be exclusively in English, almost certainly have never seen the French article.

Recycling old things written in unknown languages is, of course, not the only purpose of a book. What makes it attractive to write a book is the space which books offer. Unlike when writing an article, one is not constantly haunted by word limits. Short versions of parts of some chapters of this book have been published elsewhere. There they had to be truncated and made to fit a given context. Here we take these things up again, offer the full analysis and put them into a wider context of our own choice. A book is the form in which one can say what one has to say with a minimum of restrictions. This book combines older and newer texts in a new synthesis and develops its argument fully. That is what is new about it. Where some parts of the text have an ancestry which goes back to other texts that will be clear from the footnotes and references.

Figure 0.7 Qur'an verses, Gurbaan near Dinsoor, Somalia. A close-up of the wooden board (*looxa*, from Arabic *lauha*) on which the Qur'an verses are written. The ink, made of charcoal, can be washed off. *(Photo © G. Schlee 2005)*

EGALITARIAN PASTORALISTS

The egalitarian character of East African pastoral societies has repeatedly been stressed. The present writer in his book on the social system of the Rendille (Schlee 1979) underlines that there is no single centre of power in Rendille society and that ritual and political powers, both narrowly intertwined, are divided between different clans in a way that enforces consultation and cooperation. Harold K. Schneider (1979) has taken up the theme of pastoral egalitarianism in a wider cooperative perspective and has thereby provoked criticism by Pierre Bonte who transformed this widespread assumption into a question – 'Les éleveurs d'Afrique de l'Est sont-ils égalitaires?'[18] – thus submitting it to our reconsideration. Without denying the existence of an egalitarian ideology and social pressures towards the achievement of equality, Bonte, on the basis of works on northern Kenya and southern Ethiopia[19] complements this view by showing a number of factors which work in the direction of economic inequality and social dominance. Age-class systems, for example, can be seen as egalitarian institutions in the intra-class perspective and as instruments of separating, ranking and, often enough, dominating and controlling in the inter-class perspective. With reference to gender, it may be difficult to ascertain the degree of dominance males exert over females because labour and the other spheres of life are so clearly divided along sexual lines that very little competition or friction arises. I do, however,

[18] This is the title of Bonte (1981), translated as "Are the pastoralists of East Africa egalitarian?"
[19] See Almagor (1978); Baxter and Almagor (eds) (1978).

not know of any author who has claimed that East African pastoral societies are egalitarian also with respect to women's rights. At least in the current European understanding of gender relations, to be equal means to be allowed to do the same things and just to play separate and complementary roles, even if both the male and female roles may involve respect and enjoy recognition.

Economically, pastoralism is a matter of fortune. The English term fortune goes back to the name of a capricious goddess with changing moods. The Rendille use a different image: wealth in livestock is like the shadow in the morning and the shadow in the evening.[20] The shadow of a shade tree points in different directions in the course of a day. Impoverished nomads often get a second chance. The system of pastoral production can re-absorb drop-outs to some extent. There are loans of animals, gifts of animals, or animals as payment for herding labour which can help pastoralists whose herd-size has dropped below the level of viability to re-stock (Khazanov and Schlee, forthcoming). But there is no doubt that pastoralist societies also shed off parts of their population. Some former pastoralists take up farming in the (often already crowded) agricultural areas, others diversify and spend parts of the year in paid urban employment, often as watchmen, because their courage and fighting skills are respected and the lack of links to the urban population surrounding them may be regarded as an advantage by their employers.

If the question of whether or not East African pastoralists are egalitarian is difficult to answer within the framework of any one society,[21] the picture becomes even more complex if we include forms of interethnic association in our analysis.

Our area of study has always been occupied by a polyethnic society and not just by a number of discrete societies which lived alongside each other without interacting in other than hostile ways. Within this polyethnic society, or, as we may also say, society of societies, we can distinguish different forms of interaction between pastoral groups. Some of these forms can be called horizontal or egalitarian, like interethnic clan brotherhoods, while others can clearly be shown to be vertical and involve sub- and super-ordination. In northern Kenya and southern Ethiopia we find, however, no case where we can speak of statehood based on a pastoral system of power as, e.g. in the interlacustrine kingdoms. If we leave aside the pastoral-pastoral interaction and focus on the relation-

---

[20] For the Rendille original see Schlee and Sahado (2002 p. 82).

[21] Since this was written (see Schlee 1984b), many people have answered this question in the negative. Pastoralist egalitarianism may be limited to those who manage to remain pastoralists. Some poor pastoralists may succeed in this with the help of others. As mentioned above, redistribution mechanisms include animal loans and animals given as reward for herding labour (cf. Khazanov and Schlee (eds), forthcoming). But below a certain threshold, the poor are sloughed off: they die, settle, and/or are ethnically excluded (Iliffe 1987; Broch-Due 1999; Adano and Witsenburg 2005, 2008). Also our findings about 'Acceptance and Rejections of Christianity and Islam' (Schlee with Shongolo 2012 Chapter 3) describe some discriminatory tendencies. In the present chapter, however, we shall look at inequalities at the interethnic, not the intra-ethnic level.

ship between pastoral and other societies, then however we find that statehood is an important factor in the recent history of the area. At the beginning of the twentieth century two empires, the Ethiopian and the British, closed in on our area of study in a pincer-like grip. The formerly state-free area between these two centres of power was divided between them and new forms of interethnic dominance evolved.

This leads us to the theme of the 'nomad and the state' or 'le nomade et le commissaire', as Monod puts it,[22] and this theme is as old as statehood itself. Much of this volume will be about nomad–state interaction. But to situate this theme, let us first give a summary of the situation before the arrival of modern statehood.

POLYETHNIC PASTORAL SYSTEMS

Since the sixteenth century the Oromo established their hegemony over a large number of Semitic and Cushitic speaking peoples, with the effect that Oromo today has by far the largest speaker community of all Cushitic languages.[23] While the Ethiopian Oromo have diversified into a variety of sub-ethnicities (Arsi, Guji, and those associated with different central- ized polities in Central, Northern and Western Ethiopia), only two major groups are found in Kenya: the Tana Orma and the Boran. Oromo, in the wider sense of speakers of the Oromo language, also include a variety of groups of Proto-Rendille-Somali (PRS) origin which can be found in northern Kenya and southern Ethiopia. Of these only the Boran have been standing at the apex of a polyethnic hierarchy. The bottom of this pyramid also covers other Ethiopian peoples.

If we restrict ourselves to what today is northern Kenya, we find that the peoples over whom this hegemony extended are of PRS origin. Under the heading '*Pax Borana*' (Schlee with Shongolo 2012) we have examined how this hegemony was established and which obligations it involved for the dominant and the dominated.

So as not to repeat ourselves too much, we here only summarize the relationships these various PRS peoples had to the Boran, so that we can compare them to the interethnic relationships in the colonial period.

The associated peoples, either as entire units or lineage by lineage, were affiliated to different Boran clans, a relationship called *tiriso*, not unlike an adoption. Peoples who had a *tiriso* relationship with the Boran as a rule were sheltered against raids by the warrior expeditions, *raab*, which, in accordance with the *gada* (generation set) cycle, were undertaken once in eight years. At regular intervals delegations were sent to the *qallu* (ritual

---

[22] Monod (1975 pp. 55–60).

[23] Until 1991, Somali, though spoken by fewer people, had the advantage of being a national language and being favoured by an active language policy, while Oromo just was a minority language in Ethiopia (as Somali was in the Ethiopian context). In 1991, the Somali state ceased to exist as a functioning entity. In Ethiopia a regime based on ethnic units as building blocks has emerged and Oromo has gained recognition as a language (as has Somali, along with many other languages). In Kenya, neither the Oromo nor the Somali language has any official status.

leader) of their respective patron moiety (or kinship grouping) among the Boran with ritual gifts. Such journeys are called *muuda*. A strong blessing was believed to emanate from this practice. Office holders (*jallab, hayyu*) in the different generation-class systems of the associated peoples derived their legitimacy from the fact that – at least theoretically – they were appointed by their Boran *qallu* who sent them fur rings, *med'ic* (to be worn as bracelets), from a sacrificial animal. In practice, the decision as to whom such a bracelet was given was often taken locally. The Boran and their various associate peoples, collectively known as *Worr Libin*, formed military alliances as was effectively demonstrated by the Orma and Laikipia wars of the nineteenth century and with less success, in spite or because of the British joining their ranks, in the conflicts with various Somali groups in the twentieth century. The fact that the *Worr Libin* alliance has been stable and internally peaceful for such a long time may largely be due to the fact that the Boran, being cattle herders and warriors mounted on horses, were adapted to a different ecological niche than their camel-keeping Somaloid associates, so that competition was reduced to a minimum.

This entire network is clearly centralized. The Boran *qallus*, mainly the one of Karrayyu, were in the centre and the associates at the periphery. In many ways the peoples at the periphery, however, were ritually self-sufficient. They had their own centres, the *yaa* settlements in the case of the Gabra (Miigo and Malbe), and their own generation set (*gada*) systems.[24] The relationship between the associate peoples at the periphery was comparatively weak, although the interethnic clan relationships which are the main topic of an earlier book (Schlee 1989a) provided channels of interaction. The holy status of the *qallu* and the awe with which he was met also show that this association of peoples had a strong vertical element: the Boran towered above all others.

This may be difficult to reconcile with the thesis of pastoral egalitarianism. In the *tiriso* relationship, egalitarianism may have lingered on as an ideology – *tiriso* being described as a brotherly relationship of mutual help – while the reality was clearly one of Boran hegemony. Material profits which the Boran obtained through their position may, however, if we compare the *Worr Libin* alliance with feudal empires or modern capitalist colonialism, have been rather modest: the gifts the *qallu* obtained were of ritual, but hardly of economic importance. A higher material gain, but difficult to quantify, may be represented by the fact that the alliance saved the Boran military expenses[25] and sheltered them from enemy raiders, safeguarding their herds and territory.

Another striking feature of the *Worr Libin* alliance if we compare it to later periods is the absence of violence in enforcing internal norms. I have

[24] The extent to which *gada* systems interact through ritual exchange, thought to be indispensable, and therefore dependent on each other, has been discussed in Schlee (1998b).

[25] As arms on the level of technology with which we are dealing in this context do not represent a high value, such expenses have to be calculated as the production costs of human and equine lives, losses due to labour shortages caused by temporary absence of herders, other such disruptions of pastoral production and further costs of warfare of this kind.

heard of only one case of capital punishment where the Boran punished a disloyal Somali ally by cutting the throat of his son – in imitation of a ritual slaughter – in front of his hut. Otherwise, killing was limited to warfare, mostly in the form of raids.

Like *tiriso* also other forms of integration took the form of pseudo-kinship. Elsewhere we have discussed the *sheegat* (Som.) relationship (*himacu* in Boran means the same) by which one lineage adopted another.[26] The network of *tiriso* thus was widened to include second and third degree dependents.

It is clear that the clients or associates in such a relationship cannot be termed vassals, as is done in some of the older literature, e.g. Haberland (1963 p. 141). The similarity to European feudalism is too remote to justify use of that term. It is equally clear that the *Worr Libin* alliance lacked many features which define a state: fixed territorial boundaries, full time bureaucrats, a monopoly of violence (*Gewaltmonopol*). State control, in the form it was introduced here under colonial rule, did not improve the lot of the *Worr Libin* and may be seen as a factor harmful to them.

The introduction of tribal grazing areas, often following the course of District boundaries, was often also a disadvantage to the loyal subjects of the British. Those who obeyed these rules gave up grazing beyond the limits assigned to them without being able to restrict the access of others to their own grazing areas or being sufficiently assisted by the British to do so. Those who did not obey profited in two ways: first, their more obedient neighbours refrained from encroaching on them; second, they still managed to encroach on the resources of their neighbours. As is the case so often when policies are not thought all the way through or properly enforced, the setting up of rules favoured those who broke them. In Rational Choice theory this is known as the 'free rider problem'. The long period of internally peaceful coexistence of people of diverse origin in the *Worr Libin* alliance, and the later conflict in the colonial and, with increasing violence, the postcolonial periods, can be analysed from an ecological perspective. Specialization into different ecological niches was a way to reduce conflict and an instrument of interethnic integration. The Boran were cattle pastoralists while most of the *Worr Dasse* under their hegemony specialized on camels and small ruminants. With the recent diversification of economic activities of each single group the dividing lines of these ecological niches are blurred, economic competition is on the increase and the conditions of intensified conflict are a given.

For a closer examination of ecology and politics in the cases of two Districts, Wajir and Mandera, the reader may consult the chapter on 'Ecology and Politics' in *Islam and Ethnicity* (Schlee with Shongolo 2012). The degree to which ecological niches are kept apart or to which these different specializations blur and competition is on the increase is a prominent factor in explaining recent conflicts in the area. These are the subjects of the following chapters.

[26] See Schlee (1989a pp. 27, 29, 42–3, 46, 48, 242).

ETHNIC IDENTITIES AND FORMS OF TERRITORIALITY

This summary of northern Kenyan interethnic relations – focusing on the degree to which they were egalitarian – is based on relatively secure knowledge derived from a synopsis of different kinds of sources. Here we have only summarized earlier volumes and articles.

This synopsis has allowed us to reconstruct quite a bit of the history of local ethnic identities. It allows us to say that many of them spoke another language than Oromo before the Oromo expansion of the sixteenth century and we can specify a complex of cultural features, the Proto-Rendille-Somali (PRS) complex which dates back to that period and which they still share. Present day ethnic identifications (Gabra versus Garre, Rendille versus all Oromo-speaking camel herding groups) date back to the Oromo expansion and the different arrangement people found with it or did not find.

Later transethnic re-affiliations involved giving up one well-established ethnic identity and acquiring another. Within the interethnic clan relationships which are the subject of *Identities on the Move* (Schlee 1989a), in many cases we can distinguish those which pre-date the split between the 'modern' (post-sixteenth century) ethnic groups, i.e. those where the fissions which resulted in the new ethnic groups cut across clans, and those which resulted from later re-affiliation of individuals or groups from one already established group to another. On the borders of the lowland east Cushitic domain with eastern Nilotes, namely between the Rendille and the Maa-speaking Samburu, we also have examples of the strategic use of two well-established ethnic identities which are both maintained over time (though some might drop one in favour of the other). These people who have a reasoned claim to both Rendille and Samburu identity are the Ariaal (Spencer 1973). This is not a sign of fluidity or ill-defined identities. On the contrary, such double-identity politics are only possible when the two identities have well-defined markers and imply different kinds of rights and opportunities.

Homewood's summary of ethnic identities and interethnic relations in northern Kenya does not capture this situation at all. She writes:

> Different groups currently recognized as discrete East African tribes – Maasai (Spear and Waller, 1993), Samburu, Turkana (Lamphear, 1993), and Rendille – cannot be traced back in time as separate and distinct entities. They are current manifestations of continuously evolving identities, with different subgroups and different combinations of cultural, linguistic and economic traits diverging and others merging, adopted, abandoned and reshaped for social, political and economic reasons. (Homewood 2008 p. 38)

If the second sentence ('They are current manifestations of continuously evolving identities...') is meant to state anything more than the truism that ethnic identities are social constructs, that they are negotiated and subject to different rates of change over time, it is wrong. Precisely because for some of the groups she mentions we have relatively sophisticated reconstructions of their histories and evidence that they are not subject to much recent redefinition on the ethnic level.

Also, the authors whom she cites in support of her position do not profess nearly the same degree as agnosticism on questions of the time depth of the present ethnic identities and their relationships to each other (cf. Spear and Waller 1993). This leaves us to wonder how Homewood arrived at this statement.

When the British encountered the Gabra, Sakuye, Rendille, Boran, etc. at the beginning of the twentieth century, these groups had been in existence for some time in very similar shapes in terms of defining characteristics (group specific cultural features, including language) and composition, i.e. the clans and lineages they comprised. Later re-affiliations of sub-units took place between these well-defined major-units. The different ethnic groups have preserved their defining characteristics since in addition to (partly or as a whole) acquiring new identifications which were offered to them or imposed on them with the advance of colonialism and Western modernity (Christian churches, purified versions of Islam, affiliations to larger 'nation' states).

To speak of relative stability of these ethnic identities does not mean to espouse primordialism or even just to claim that ethnic identities generally have a high resistance to change. I regard these matters as a purely empirical question: social identities can be observed to have higher or lower speeds of change in comparison to each other or in comparison to earlier or later versions of themselves in different periods. In northern Kenya I have found many clans to be older than the ethnic groups which comprise them, because the ethnic splits occurred across still existing clan identities which must already have been in existence for the split to go right through them. On the other hand Newbury (1980, cf. Schlee 1989a p. 136) has found out that the Kinyarwanda speakers in what is now Rwanda and Congo do not share any clans. They split from each other about 200 years ago and all their clans must have evolved since. The conditions under which clan identities change faster than ethnic identities or vice versa must be looked for case by case, and the same applies to all other social identities. We know that their speed of change sometimes accelerates and sometimes slows down, but the 'why' question is under-researched. For too long the study of social identities has been caught in sterile dichotomies like primordialism versus instrumentalism (often wrongly implying that either there is no change at all, at least no intentional change, or that change is ad hoc and arbitrary and can take place any time). Identity change, can, of course, mean many things: people can drop one identity and adopt another, or a given identity itself can change by being re-evaluated, re-defined or politicized in a number of ways. This, however, is not the place to go deeper into the theory of collective identification (cf. Schlee 2008a, Donahoe et al. 2009).

The British did not 'invent' the ethnic groups of northern Kenya. That ethnic groups were a colonial invention may be valid for some other parts of Africa (e.g. Lentz 2006) but not for northern Kenya. But that they did not invent ethnicity does not mean that it did not change its character under colonial influence. It did: ethnicity was territorialized.

'Territorialized ethnicity' refers to the same phenomenon as 'ethnic territoriality'. The two concepts do not differ in their denotation but in their derivation. An ethnic group is territorialized and as a result of that process territorial, in contrast to those groups whose identity is not tied to a particular piece of land, but only to a professional specialization, a genealogy, a religious affiliation, or the many other things which can exist without a fixed locality or a bounded surface[27] and can serve as social identification. The contrast here is between territorial and not territorial. 'Ethnic territoriality', however, starts with the presence of a territory or territories and contrasts different forms of identification, which justify a claim to such a territory. Ethnic territoriality stands in conceptual contrast to (but, of course, may be combined with) individual ownership, the nation-state, and the emotional and ideological attachments to the Fatherland that come along with it, combinations of personal rule with religious affiliation (cf. *cuius regio, eius religio* – the famous principle of the Westphalian Peace) and many other relationships between people and land, which are of a possessive and potentially exclusive nature and therefore justify the term territorial.

Ethnicity as such has no territorial implications, and, in this regard, the concept of an 'ethnic group' differs from that of a 'nation'. To say that the Kurds are the world's largest nation without a state implies a problem. The implication is either that they should have a state or, as some fear, that they might succeed in getting one. Nationhood implies the claim to (territorial) statehood or at least to regional autonomy within a state as the next best thing.

With ethnic groups it is different. Where we find professional specialization along ethnic lines, as in West Africa, ethnicity can only have limited territorial implications or none at all. Professional specialization implies having to meet on the market place. It cannot be combined with a high level of territorial separation. And there are other examples of ethnic groups, defined with reference to other shared features, that do not possess territories or claim them. So, instead of a correlation or close dialectic between ethnicity and territoriality, we find variation in the ways in which and the extent to which they go together.

In this sense, the relationship between ethnicity and territoriality resembles that between ethnicity and language, which has been summarized elsewhere (Schlee 2008a pp. 99–103). At one extreme, there are cases in which the members of an ethnic group, well defined by many other features, do not share a language, and, at the other extreme, there are ethnic groups that are defined almost exclusively by their language. In between, there are many other possible relationships between language and ethnicity. The same kind of conceptual logic applies to the relationship between ethnicity and territoriality.

---

[27] See Schlee (2008a) for a systematic treatment of forms of identification and the conditions under which they take place.

THE EMERGENCE OF TERRITORIALIZED ETHNICITY
IN NORTHERN KENYA

In northern Kenya, territorialized ethnicity and thereby ethnic territoriality have emerged in their present form in the colonial period and have been greatly politicized since. The British order was a territorial order. After set-backs during World War I, when British forces were needed more urgently in Tanganyika (now Tanzania) to fight the Germans, the British managed to establish a degree of territorial control in northern Kenya. They delineated territories for each of the pastoral groups, as they perceived them as such. Soon the tracks, which led from different sides to Wajir, and a number of cut lines in the vegetation served as boundaries between 'tribal grazing areas'. If a herd was found on the wrong side of a line, ten per cent of it was taken as a fine. To the pastoralists this did not make much sense, except that some of them thought that these ten per cent constituted the livelihood of the British who, like all of us, need to live off something.

This form of control based on bounded surface areas and lines on the map differed greatly from earlier systems. Before the British, the Boran had established hegemony[28] over much of northern Kenya. Many groups of lowland camel pastoralists, who spoke Somali-like dialects or had spoken such dialects before they adopted the Boran variety of Oromo as their form of speech, brought them regular presents to their ritual centres in what now is Ethiopia, from an economic perspective a very light burden, and received a blessing from the *qallu*, the ritual head of one or the other of the two moieties to which the Boran and all of their allies were associated. Also the age-grading (*gada*) systems of some groups took chronological clues from each other and involved ritual exchange (Schlee 1998b). Before the Boran influence, these groups of Lowland East Cushitic[29] speakers had their own, independent *gada* type generation set systems that were – along with many camel-oriented rituals and a specific calendar – part of an earlier Proto-Rendille-Somali (PRS) complex of cultural features (Schlee 1989a). This picture of social relations is made up of both difference (Somali/Somaloid/Oromo speakers, cattle/camel husbandry, distinction along interethnic hierarchy) and interaction (co-residence in the same or adjacent areas, sharing of water points, economic exchange and ritual interdependence). It was a system organized along the lines of differences without separation. This interethnic system, the Boran-centred alliance known as *Worr Libin* (People of Libin), also had a military aspect.

[28] Some authors (e.g. Kassam 2006) have taken the term hegemony to imply an accusation against the Boran and have regarded it as offensive. I have quite deliberately chosen this term rather than domination or colonialism, because it implies a superior status among others who, to some extent, are equals and situationally regarded as such, and it is not based on direct administration or the constant use of violence. I still find it quite appropriate to describe the relationship between the Boran and the other groups of the *Worr Libin* alliance, implying neither too much nor too little inequality (Schlee 2008b).

[29] Lowland East Cushitic languages comprise Rendille and other Somali languages (Maxatiri, Maymay), Arbore, Afar, Saho as well as Oromo.

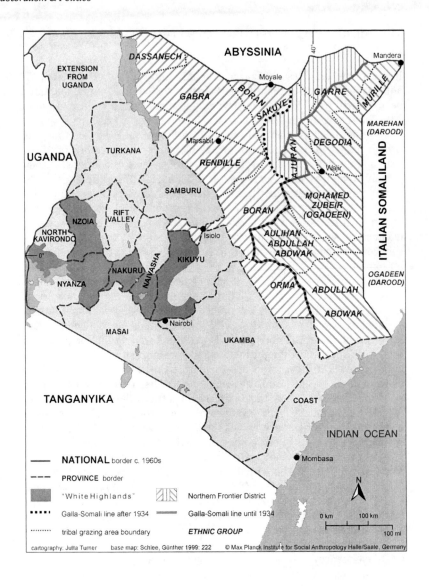

Map 0.1 Boundaries in colonial Kenya (based on Schlee (1999))

The Boran, jointly with their camel-keeping allies, repelled the Laikipiak Maasai who had ventured far into northern Kenya.

The Laikipiak Maasai are well known from the accounts of the Maasai civil war by early European travellers. These Laikipiak scattered after their defeat by the other Maasai (Thomson 1968 [1885]), and a large body of their warriors moved north from what is now the Laikipia District,[30] took the whole of Rendille, people and livestock, as their spoil and divided them up among themselves, leading a brief and happy life for as long as their superior force enabled them to do. However, when they wanted to move the whole Rendille society elsewhere, the Rendille warriors who had been hiding in the bush chased them away in a bloody battle.[31] The Laikipiak then moved north, raiding cattle and driving large herds along, until they were beaten by the *Worr Libin* cavalry near Buna.[32] A British compiler[33] of the accounts of 'some old men' gives 1876 as the probable date of the battle and Korondile as its place.[34] Whatever the exact location might have been, it is clear that these Maasai had ventured far into the Boran heartland. The father of our informant Waako D'iriba[35] took part in this battle, and Ido Robleh, the Ajuran leader, is also reported to have borne 'the mark of an arrow got near Buna in one fight with them'.[36] Apart from the Boran and the Ajuran, the other member peoples of the *Worr Libin*, the Garre and Gabra,[37] were also involved in these fights and the subsequent pursuit of the intruders, and temporary solidarity was extended even to the Rendille – who stood outside this alliance.[38] The only people who did not join the Boran in these fights were Warra Daya, who had had a bloody conflict with the other *Worr Libin* in the preceding decades and had withdrawn to the south[39] where the Daarood Somali were to continue their decimation. Their remnants can be found among the Tana Orma.

As far as the Somaloid associate peoples of the *Worr Libin* are concerned, we can say that they were strongly engaged on the side of the Boran in the Laikipiak war, as they had been earlier in the Boran/Warra Daya conflict. The wars of the nineteenth century thus show that the alliance

[30] The District capital being Nanyuki, 200 km north of Nairobi.

[31] Common Rendille tradition. Cf. Grum (1976), Schlee and Sahado (2002 p. 113).

[32] Conversation with Waako D'iriba, Boran, Marsabit, April 1980.

[33] Kenyan National Archives, *Moyale Station Report*, microfilm Reel No. 43, 1917.

[34] Paul Robinson (1985) has collected the names of years used by the Gabra. Here we find the Wednesday year 'Arba Gabbri Urbur dat', explained as the withdrawal of the Gabra from the Laikipiak northwards to a place called Urbur, for 1879 and 'Kamis Kibiyi basan' the Thursday year, when the Laikipiak were chased, for 1880.

[35] Conversation with Waako D'iriba, Boran, Marsabit, April 1980.

[36] 'Notes on the traditional history of Wajir tribes,' *Wajir Political Record Book* Vol. II, Kenyan National Archives, compiled 1939 from accounts by Ido Robleh, Dima Abdi, other elders and from other sources. The 'Korei' (Korre = Maasai and related peoples) here are not specified as Laikipiak.

[37] No information on Sakuye participation.

[38] Conversation with Boru Galgallo, Odoola, Gabra.

[39] Kenyan National Archives, *Wajir Political Record Book* Vol. II, 1939; Schlee 1992b.

Figure 0.8 Water well near Melbana, southern Ethiopia, 1986 Near Melbana in southern Ethiopia there is a cluster of wells which form part of the 'nine wells', *tula saglaan*, of the Boran core area. Access is via a long dug-out ramp leading to a lower level where it is wide enough for a herd to gather and where earthen troughs are built. Water is passed along a chain of men in a well from the water table which is yet much lower and poured into these troughs. Here the trough has been filled for a herd of cows but horses, which roam about freely, have been faster and finish the first fill. Horses, by Boran customary law, cannot be denied water because of their former importance for warfare and for transmitting messages. *(Photo © G. Schlee 1986)*

was then strong and functioning.[40] We have had a closer look at these wars to ascertain that there was an accumulation of power, that there was an organization, that there was ritual and military cooperation in this pre-colonial pluriethnic society. All this was achieved without formalized districts and provinces.

How then were the *Worr Libin* organized? All we know is that they did not have boundaries delineating surface areas. Even the words for 'boundary' in the local languages are loanwords from other languages (Schlee 1990b, 1994c). So what do we know about the spatial organization of this cluster of peoples in the absence of boundaries? In the wet season, people and herds must have dispersed and mixed widely, but in the dry season the Boran certainly exercised some control over the access to pastures by controlling wells. No thirsty herd could be refused water once, but the people in control of the water management (*herega*), the 'father of the well' (*abba eela*) and those who acted on his behalf, could certainly arrange the watering schedule in a way that made it unattractive for

[40] As this is the only point I wanted to illustrate here, I refrain from unfolding the rich and vivid traditions about the Laikipiak war, which have been collected from all over northern Kenya.

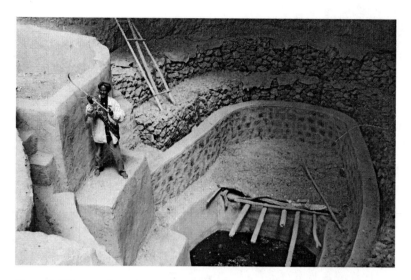

Figure 0.9 Water well near Melbana, southern Ethiopia, 2010. The same well as in Figure 0.8, 24 years later. In 2010, with the help of international donors the well and trough have been cemented. Such wells, improved or unimproved, belong to clans and access to them is regulated by committees. *(Photo © G. Schlee 2010)*

strangers to come back. Another factor that influenced the distribution of people and livestock on the ground was war. War did not have territorial expansion as its aim. Stated aims were the acquisition of the honoured killer's status and taking livestock as loot. Not all pastoralist groups were members of the Boran-centred alliance. Those who were not were better off staying at a distance. Warfare may well have had territorial effects, even though its primary objectives were different.

The only form of direct control of grazing (rather than through regulating access to wells) seems to have been applied to a category of land called *kaloo* in Boran. With reference to Boran villages on the foothills of the Ethiopian highlands, where some agriculture is possible, Bassi (2005 p. 145) describes *kaloo* plots as small, 'fenced or marked with branches', 'close to villages and often adjacent to cultivated fields, and meant for "calves" in case the dry season goes on longer than expected.' The calves in question are those which belong to those few cows one keeps in the settlement for milk supply as long as conditions permit. For the far ranging *fora* or satellite herds in the care of young men there is no reserved grazing. With the advancing dry season they would move wherever conditions are still most bearable.

Leus (2006 p. 383) traces the term *kaloo* to the Guji usage of the Oromo language. The Guji are another subgroup of the Oromo, bordering the Boran in the north. The relationship between the two groups has been hostile for long periods (Tadesse Berisso 2009) and the Guji have recently expanded at the expense of the Boran. Leus explains that the 31

Figure 0.10 Degodia collecting water, Marsabit District. During a good rainy season the animals do not need to be taken to wells because they get enough moisture from the vegetation. Water for household consumption can be collected from ponds of rainwater. Here Degodia collect water from such a pond in the eastern part of Marsabit District. *(Photo © G. Schlee 1981)*

Boran adopted the *kaloo* reserves only in the *gada* period of Jaldesa Liban (1960–68), because before that there used to be abundant grass and no need of grazing reserved for calves (*kaloo yabbiyee*) or lactating cows (*kaloo haawwicha*).

The Rendille term for *kaloo* is *liig*. It refers to the area immediately around a nomadic hamlet where the kids, lambs and (camel) calves graze during the day. The young are separated from their mothers during the day to prevent them from drinking milk, so that there is more to be milked for human consumption in the evening, and to enable the adult stock to reach pastures in a wider radius. This zone is not demarcated in any way. Rendille just regard it as bad form not to respect it and to graze their animals too close to the temporary settlements of others.

Such rudiments of special rights to pasture on a small scale have nothing to do with the tribal grazing areas and ethnically based districts the British were going to introduce. Before the arrival of the British, pasture on this larger scale was a matter of open access by people who basically regarded each other as equals, and it was only indirectly controlled, if at all, by granting or withholding rights to water at the wells. It was the British who introduced maps and drew lines on them. But, as we shall see, these lines did not disappear when the British departed.

THE PROGRESS OF TERRITORIALIZED ETHNICITY IN KENYA SINCE INDEPENDENCE

Kenyan Independence in 1963 came with much rhetoric of African solidarity against the former colonial masters, nation-building,[41] and unity. Yet it was clear that the ruling party was based on earlier regional and tribal organizations and that the civil service was dominated by the Kikuyu, the tribe of the president (Kioli 2009). Both the colonial order and the African opposition within it had shaped Kenyan politics along ethnic lines, and ethnicity was well on its way to acquire its present territorial and exclusionist form. In the pastoral north, people had been misled by a referendum held by the British to believe that joining the newly independent Somalia was an option. The results of the referendum made the newly founded independent Republic of Kenya aware of the lack of 'Kenyan' feelings among some of its (so far marginalized and unnoticed) citizens, and it moved in with its armed forces. The *shifta*[42] war lasted for most of the 1960s. Free movements of people were prevented by insecurity emanating from the insurgency and the counterinsurgency. The Sakuye were concentrated in camps, guarded by the army, where they lost all of their livestock. Other groups ranged as far as they dared. Far from overcoming the colonial heritage, local chiefs have since continued to invoke old colonial boundaries in order to keep 'intruders' off.

Still, as late as the 1980s, ethnic politics were not part of the public discourse. Ethnicity figured as the interpretation of others and the hidden agenda. To accuse others of being in pursuit of ethnic or tribal interests was part of the critical discourse. That changed in the 1990s.

In the later years of President Moi's term of office, Abdullahi Shongolo and I analysed a number of conflicts between pastoralists in northern Kenya and also collected press cuttings to identify reactions by both the wider public and the government and forces going out from these external agents, which may have influenced these conflicts between pastoralists. It was quite clear to us, and remains so retrospectively, that the late 1990s and the years around 2000 in Kenya were a period in which ethnic territoriality dispensed with all sorts of excuses and came out in the open. This is the reason why we take up our historical account from that point of time in the present volume (chapter 1).

Again, local patterns of conflict among pastoralists in the north were influenced by national politics or regional politics in other parts of Kenya. In order to mobilize the government for their own causes, local leaders had to find out which degree of ethnicization had become usual and subsequently legitimate elsewhere. There the ethnicization of

---

[41] The term 'nation-building' to denote a political programme makes clear that there was no nation when Kenya was founded as an independent state. While national emancipation in Europe (as in the formation of new states after World War I in accordance with Wilson's Fourteen Points) assumed the existence of nations prior to the point of time when they achieved statehood of their own, Kenya at independence had the shell of a 'nation-state' yet to be filled with a 'nation', a process for which some people are still waiting. It is clearly a case in which the state precedes the nation.

[42] Amharic for 'bandit'.

politics and the tolerance towards or even promotion of ethnic violence proceeded in giant steps, and the idea that every group had a homeland and the right to expel minorities by force gained ground.

Here, global forces come into the picture. As long as superpower rivalry was a major factor in a bipolar world, i.e. prior to the dissolution of the Soviet Union, human rights, civic rights and democracy were not a major issue in Africa. Dictatorships and one-party systems were tolerated as long as they were viewed as dependable allies. The early 1990s, however, brought a completely new political climate, one that favoured, nay demanded, 'democratization' and multi-party systems throughout Africa. In Kenya, the Moi government introduced multi-party democracy under pressure from the International Community and the internal opposition. Moi had predicted that multi-party democracy under Kenyan conditions would lead to ethnic violence. There were good reasons to assume this, because ethnicization of politics and a culture of disregard for the law had made progress in Kenya by that point of time. Apart from that, as president of the country, Moi had all the means of making his predictions come true.

While in the 1970s the ethnic dimension of politics was unofficial and talked about like a dirty secret (although everyone was aware of it), in the 1980s ethnicity came out into the open. It had become legitimate. By the 1990s I was often astonished how openly politicians who claimed to represent a tribe also claimed territorial control of a given area in complete defiance of the existing law. Members of Parliament (MPs) threatened to prevent political rivals from visiting their constituencies. The press reported such statements without critical comments.

In the following chapter it will be shown to what degree ethnic discourses had penetrated electoral politics by the 1990s. There are cases in which MPs favoured the admission of pastoralists from elsewhere into their constituencies in exchange for their votes, and other cases in which populist and xenophobic rhetoric was adopted to keep strangers out and to gain the vote of the resident local ethnic majority. All this is based on the assumption that there are group rights to specific territories.

Appadurai (1991 p. 193) has pointed out that '[d]eterritorialization is one of the central forces of the modern world'. There is no doubt that this is true. There are, however, also counter-tendencies. The lives of pastoralists and many others in northern Kenya and southern Ethiopia are deeply affected by a process in the opposite direction: territorialization. Here the nation-state, a far-travelled global model, is applied *en miniature* to create ethnically pure local territorial entities. Ironically, this system of mutual exclusion, reminiscent of *apartheid* with the difference that it was used by Africans against other Africans, gained ground in Kenya around the time it was abolished in South Africa.

I

Moi Era Politics,
Transnational Relations and
the Territorialization of Ethnicity
ABDULLAHI A. SHONGOLO & GÜNTHER SCHLEE

People of the Degodia and Ajuran Somali clans are culturally indistinguishable at first sight[1]. There are some Boran-speaking Ajuran, but the remainder of them speak the same form of Somali as the Degodia. An elderly woman, Ajuran married to Degodia, once explained that the division of labour between the genders differs between the two groups, and that it was a bad surprise for her to find that as a married woman among the Degodia she had to herd animals. But such differences only turn up when the respective inquiries are made or one has spent more time living with these clans. When Schlee visited Somali hamlets in the eastern part of Marsabit District and in Wajir District in 1977–1980, he found Degodia and Ajuran living in intermingled nomadic hamlets and there was no way of telling which was which without asking.

When Schlee met the same people again in 1984, one of his key Degodia informants, a very hospitable and wealthy camel owner, had been robbed of all his camels by Ajuran and was collecting money to take the matter to court. This remained without success because the same had happened to thousands of others and the courts did not interfere in matters relating to this 'war'. Some Ajuran informants, no less helpful and generous to Schlee, boasted that it was 'them' (maybe not individually but people close to them about whom they knew) who had taken the camels of his Degodia friend. Of course, Schlee could not make any use of this information to help his friend, because anthropologists, like journalists, have to protect those who give them information.

In the meantime mutual raiding had occurred between the groups and herd girls had been raped. There was even talk about girls who had been ripped up after being gang-raped. The Ajuran had better links to the Government. Rumours have it that the head of the Army, an Abud Wak Somali from Garissa, sympathized with the Ajuran because his people were having similar problems with the Degodia at the other end. Thus the notorious Wagalla massacre of 1984 came about. Thousands of Degodia were rounded up on an airstrip near Wajir where they perished by hundreds in the sun.

---

[1] Earlier shorter versions of this chapter, each singling out a different aspect, have been published as Schlee (2007) and Schlee (2009a, 2009b). This is the first occasion where space is given to combine the whole interethnic mosaic in a broader picture.

1984 was a very dry year and pressure on pastoralist resources was high. Aulihan Somali from Garissa District pressed into Isiolo District and started to squeeze the Waso Boran out of their pasturelands. Many Boran herds were raided. Also Degodia had moved into Isiolo District, partly because of the drought and partly because of the Ajuran threat, but they managed to establish a better relationship with the local Boran. They helped them to retaliate against the Aulihan. That they helped their traditional enemies, the Boran, against their fellow Somali, is not surprising if one considers what they had suffered at the hands of the latter.

As a consequence, Waso Boran informed the Boran of Marsabit and Moyale that the Degodia should henceforth be regarded as allies. The message, on which a committee of elders had agreed, was taken by Isiolo Boran who took herds to the Moyale area to known Moyale elders. In that year there was much coming and going between Moyale and its Ethiopian hinterland, and the Waso. Guns were purchased in Ethiopia to be used against the Aulihan or for re-sale. Degodia then flocked into the Moyale area which at that time still belonged to Marsabit District (Moyale became a separate District in 1995). The pasture was sufficient for their camels which can make use of a much wider radius around the water points than the cows of the Boran whose carcasses at that time were lining the roadsides. For them the walks between pasture and water had become too long. Collecting grass from remote mountain sites inaccessible to cows had become a business for women around Moyale town in that period and the price of a cow was equivalent of the price of its fodder for three days. In the Moyale market, the milk of Degodia camels had replaced that of Boran cows. Through all ups and downs in the Boran-Degodia relationships some townspeople were always to remain opposed to expulsions of Degodia because their camels guaranteed a reliable milk supply, mainly used for the ubiquitous tea.

Some Degodia, who had picked up bits of the Boran language already in the Waso area, started now to identify themselves as Boran when asked for their tribe.

> '*Att gosi maan?*' '*Booran.*' – 'Which tribe are you?' 'Boran.'
> '*Balbal tam?*' '*Degodia.*' – 'Which section ('gate')?' 'Degodia.'

The Degodia thus had become, at least in some situations, a section of their traditional enemies. In one of the ad hoc genealogies which sprout in such situations, the Degodia were not depicted as descending from the Boran but as branch collateral to them. Both 'Degodia' and 'Boran', treated as eponymous ancestors, were said to be the sons of one ᶜAli. The Boran tolerated this usage and took it as a sign of good will, although it was contradicted by other such genealogies, like the one attributing to 'Boran' and 'Waat' (the putative ancestor of the former hunter-gatherers) a shared father 'Horro'. The Gabra, who had a long-standing *tiriso* relationship with the Boran without ever having claimed common ancestry with them, were amused by this genealogical fancy.

36

Nothing derives from its roots. A plant grows from the surface, where its seed has come to lie, both ways, up and down at the same time. When the root metaphor is used in the context of human history, it is mostly used wrongly. Peoples and states are said to stem from their roots. In reality, they drive their roots from the present into the past, from the surface of time into its depth, just like a plant drives its roots downwards into the soil. The Degodia had just been trying to grow roots in a new soil, as Somali groups and other lineage-based societies have always done, by linking their genealogies to some real or imagined ancestral figures of the local society. Had this genealogical fiction persisted long enough, people might have started to believe it.

The Degodia gave some substance to their new relationship with the Boran by guaranteeing them safe passage from Moyale to Isiolo in a straight line. The road diverts from this line far to the west and had become obstructed by so many police barriers where all lorries, with or without insufficient documentation or other faults, had to pay bribes, so that road transport had become very costly. Boran traders, and Burji traders who were not distinguishable from them, took advantage of the Degodia guarantee of security and trekked herds by foot to the central Kenyan markets by the straight route through the bush. This route came to be known as the 'Bosnia route'. In this period 'ethnic cleansing' was frequently heard of on the transistor radios, which formed the link of these mobile populations to the outside world. This new phrase was used in connection with Bosnia. As the new route had been cleared of Ajuran, it was believed to conform to Bosnian standards of ethnic un-mixing.

The Degodia did not distinguish between Boran and Burji traders, but the Boran traders did. The Boran used to despise the Burji, who were originally mountain farmers and poor in cattle-wealth, and the new role of Burji as highly successful competitors in business did not endear them to the Boran either. So the Boran sacrificed a billy goat (*korbes*) at a place called Mad'eera Kaayo along this route to prevent the evil associated with Burji from coming along with them on this route. The Burji were not informed of this.

In the same measure as the relationship between Degodia and Boran improved, the one between Garre and Boran declined. In a way the Degodia were occupying the niche formerly occupied by the Garre as camel pastoralists in the Boran orbit. In the Negelle area in Ethiopia raids occurred between Garre and Boran, and also Degodia took advantage of this situation for raiding the Garre there. These were not the same Degodia as those who had come from Isiolo District to Moyale, but people of the same Somali clan who had been living in Ethiopia for some time. Cattle taken as loot in these raids were taken to Kenyan markets. The Kenyan Garre were now convinced that the cattle taken by Moyale traders by lorry to Nairobi rightfully belonged to Garre. Influential Garre then succeeded in influencing the Veterinary Department to close the road to cattle transports under the pretext of the outbreak of some 37 disease somewhere. These transport restrictions 'for veterinary reasons' have a long tradition of being misused for political and economic purposes (Schlee 1990b, 1998a). The closure of the road also contributed in diverting the trade to the 'Bosnia route' and thus to strengthening the interdependence of Boran and Degodia to the detriment of the Garre.

In the second part of the 1990s, however, also the relationship between Boran and Degodia in what by then was the Moyale District was worn thin. Dr Gurrach Boru Galgallo, in his 'Manifesto for the total development of Moyale Constituency' which was part of his successful election campaign for a parliamentary seat in 1997, depicted the situation thus:

> Also, our neighbours from North Eastern Province have penetrated deep inside our land and their camels roam every inch of our soil. These camels have caused serious environmental degradation by destroying our vegetation. Further, their elephant-style guzzling of water in our small dams has led to perennial water shortages both for our livestock and human consumption. Our leaders have simply turned a blind eye or even facilitated this naked aggression of our territory because of personal gains. (p. 9)

> In addition to depleting our resources, the intruders from NEP [North Eastern Province] alluded to above, also engage in cattle rustling and other banditry activities. These people have acquired guns from their kins [sic] in Somalia following the collapse of the Barre regime [January 1991]. At that time, there was an influx into Kenya of armed refugees and gangsters from the various fighting sections. Guns were cheaply purchased and sometimes even exchanged for food. (p. 11)

We have decided to let this chapter be multi-vocal. Instead of trying to extract the bare factual information, we allow the different views to be represented by original voices. It goes without saying, that this manifesto is value-laden and aims at the political mobilization of Boran against Garre. There will be other types of texts which we will cite in the original or in a close translation: the statements of elders, their transformations into the 'minutes' of a meeting, and press articles. All these kinds of texts have their particular biases and blind spots. As conflicts are not only fought with arms but with words, we decided to leave as much original wording of the different views about the conflict as possible, also to show some of the dynamics of verbal escalation.

A new factor which came to influence both internal Boran politics and their relationship to neighbouring groups was the presence of the Oromo Liberation Front (OLF) in the area. The OLF, which had been operating in Ethiopia for decades, had to withdraw more and more to the Kenyan side of the border since 1992. A Gumi Gaayo Assembly (GGA), a huge gathering of Boran which takes place once in eight years, took place in 1988 and was attended by Abdullahi Shongolo (for his account of an earlier such pan-Boran legislative meeting see Shongolo 1994). It was used by Ethiopian Government representatives, among them the President himself, Dr Nagasso Gidada, an Oromo, to appeal to the Boran to withdraw their assistance from the OLF. The GGA then decided, according to the proceedings written up in English by Goolo Huqa:

> The OLF fighters are called to come back by GGA and resume their peaceful life.
>
> It was agreed by both the GGA and the Government that there will not be any measure to be taken against OLF members who may come back. They shall not be accused of any wrong-doing they might have done during the last five years in connection with the liberation struggle. The GGA advised all the Boran to reinforce the decision. Everyone living in Borana territory is obliged to obey the GGA decision.
>
> The committee consisting of members from Dirree, Moyyaale, Yaavalloo, Areero, the gada leaders, prominent persons, administrators and army officers shall be established to facilitate the coming back of the OLF members. Each *warada*[2] mentioned above must raise EB [Ethiopian Birr] 2000 which shall be used by the committee while on duty. The Boorana zonal administration office is ordered to provide a vehicle when required by the committee.

The Ethiopian Government thus successfully won support of the Boran community by using the Abba Gada[3] as a government agent to propagate and execute the state policies to the Boran. The Boran traditional leadership has been fully recognized and the government is using them to reach the Boran not only in Ethiopia but also in Kenya, an area where the *gada* laws were less effective in the recent past.

In Moyale electoral politics there has always been a marked rift along the moiety line which splits the Boran: there always has been a candidate from the Sabbo moiety and another one from the Gona moiety. Both stemmed from the senior *qallu* clans: Karrayyu in the case of Sabbo, *Worr Jidda* (Jilítu) in the case of Gona. It is therefore unthinkable that the Boran agree on one candidate. The moiety rivalry dictates a minimum of two. In 1997 not less than eight Boran candidates were competing for the parliamentary seat. This gave the smaller groups like Gabra and Garre an influence in excess of their numerical proportion. The incumbent Member of Parliament who had to defend his seat in 1997 was Mohammed Galgallo from *Worr Jidda*. If he had succeeded in being elected for a second time, that would have been the first re-election of an MP in the history of that constituency, which is known for the intensity of its infighting and its rapid wear-down of office-holders. He was not re-elected in the end. Causes of his failure include the following: according to his adversaries, to strengthen his backing by the Boran, Mohammed Galgallo tried to remove non-Boran Civil Servants from the District and fill the posts with his own sympathizers. He singled out Garre for removal while Degodia office holders were not affected. He also exempted Ajuran from his attempted job discrimination. The Ajuran, according to their old *tiriso* relationship, are Jille = Jillítu *qullullu,* 'bald *Worr Jidda*', i.e. *Worr Jidda* without the tress

---

[2] Also spelled *woreda*, meaning district.
[3] The highest office holder in the Boran *gada* system who resides near Negelle.

of traditional Boran elders. Mohammed Galgallo thus had reason to treat the Ajuran as non-Boran members of his own clan. These policies, which were attributed to Mohammed Galgallo, earned him a headline in the *East African Standard*, Wednesday, 12 June 1996:

> 'Nepotism alleged. Minister slammed.'
>
> A group of Moyale District elders yesterday accused the local Member of Parliament, Mohammed Galgallo [. . .].
> The elders claimed Galgallo, who is also an Assistant Minister for Commerce and Industry, was allegedly fighting for the removal of non-Boran civil servants from the district.
> Led by the local Kanu chairman, Hajji Khalif Ibrahim [a Garre], they claimed the Minister was encouraging disunity and hatred [. . .].

He was also accused of discriminating against Karrayyu, and frustrated Karrayyu job applicants, plot registration applicants and business men who had not even received tender notices were all desperately determined that this time a Karrayyu candidate should be elected.

He had openly criticized the Government for not doing enough to improve the security situation in the Moyale District. This had even led to his removal from the position of an Assistant Minister in the Ministry of Internal Affairs and a seven month period as a mere MP before he became Assistant Minister again, this time in the Ministry of Commerce and Industry. There is no doubt that this criticism was justified and his courage laudable. The consequence of the weakness of the Government forces in the area to him personally was, however, that he came more and more to rely on the OLF, the Oromo Liberation Front, as a protecting force for the Boran and a source of political support for himself. The OLF had a strong base near Sololo and another one on Hurri Hills. His sympathies with the OLF were strengthened by the circumstance that the Ethiopian Government had been replacing '*Worr Jidda*' officers by 'Karrayyu' officers in their sphere of jurisdiction, i.e. members of the Gona moiety by people of Sabbo (in political rhetoric members of other clans of the two moieties tend to be counted with their respective dominant clans). Mohammed Galgallo's OLF inclinations made it easy for his opponents to denounce him as a disloyal Kenyan who relied on foreign assistance. That they themselves were accorded massive assistance by the Ethiopian Government, another outside force, was not put forward by them in this context.

Boran in public office in Ethiopia declared publicly that any Boran living in the area would be free to register as voters in Kenya. Unofficially they especially encouraged people to vote for the Karrayyu opponent of Mohammed Galgallo, Dr Gurrach Boru Galgallo.

In the same period as the voters' registration, the Abba Gada, who resides near Negelle, Ethiopia, held many meetings in both Ethiopia and Kenya to tell the local people to persuade the OLF fighters to give up and come back to help in the reconstruction of their home country. He had an Ethiopian Government vehicle at his disposition. This had been going on for some time. It was inevitable that Mohammed Galgallo would be involved in this discussion sooner or later.

Before and after the Gumi Gaayo Assembly in August 1996, the Abba Gada, by the name of Boru Mad'a, had a series of meetings both with Ethiopian Government officials and the Boran elders at several localities. With an arrangement between the Ethiopian and Kenyan authorities, the Abba Gada also held meetings with Kenyan Boran and the local leaders. In August 1996, just before the Gumi Gaayo Assembly, the Abba Gada held his first meeting at Butiye, a large settlement which is part of Moyale (Kenya) Town, where the majority of the residents are Boran. He met with the Boran local leaders, the chiefs, the councillors and selected elders from each sub-location in the vicinity which had a Boran majority. In his short speech he told the local leaders how he had held discussions with the *gada* leaders about the unresolved Boran and Kenyan Government attitudes towards the OLF movement. He also briefly narrated the problems the Boran community had gone through as a result of Boran support for the OLF. At the meeting, at first the Abba Gada's proposal met some resistance, but in the end an agreement was reached to remove the OLF from the area with the support of the two governments. The Abba Gada blessed the gathering and promised to visit the Moyale community from time to time to look after this matter. Four months later, in December 1996, the Abba Gada returned to Moyale with a delegate from the *gada* council. This time OLF supporters and representatives of the OLF also joined the meeting. The Abba Gada reiterated his call and asked the leaders of the area whether his message was received by the OLF. Here are excerpts from Shongolo's notes about the ensuing discussion:

Diima Diida (*hayyu*):
We have heard speculations by Moyale (Kenya) Boran that the Abba Gada had been corrupted by the Tigre Government[4] and that this is why he has been forced to call for the peaceful return of OLF. If you think that that is true, why don't you also mention those who have enriched themselves with the money given to them by the OLF? Some of them are at this meeting. You have built shops and stone houses which people have been pointing out to each other. We need not listen to speculations. Now the Abba Gada himself is here. The decision to call for the return of the OLF was made by the entire Raaba Gada and if any one of you would like to dispute the decision you are now free to do so basing your arguments on truth and only truth. One thing must be clear to you that those youths, who are in the wilderness, are neither our sons nor yours unless they abide by our decisions. Their decision to form this organization (OLF) was neither discussed with us nor with you. We do not say that it is only your responsibility to bring them home but your role, just like mine, is to request them to come home or just to pass over the message of Raaba Gada. If they listen to you and heed your call, please ask them to return home, and if they refuse to listen to you, we shall not put any blame on you.

---

[4] Meaning the EPRDF Government which evolved from the TPLF.

Golic Guyyo, a Kenyan elder, asked:
If now these youths refuse to return home, and do not listen to us just like they did not listen to us when they formed their organization, what are we going to do about it? And, secondly, supposing they respond positively to our call, you know that they have arms, how will they come home and how will they be integrated into the community?

Diima Diida:
An agreement has already been made between the two governments that should these boys come home even in daylight carrying their arms, no one will ask them or arrest them. You all know even someone like Hassan Gooro, a notorious *shifta* leader who has caused much devastation to this land, today is back with all his army and their arms and no one has questioned him. He and his men are just living among us not even feeling the guilt of their evil deeds. Likewise our youths will not be questioned for the past for the Abba Gada has given them amnesty. An arrangement has already been made to integrate them into the government armed forces. Just like the former Gabra and Garre returnees, they, too, will be integrated into the forces. Those who do not want to join the army are free to live on their own as they please. As you know the government has seized some livestock of those youths who joined the OLF. On their return, their animals will be given back to them. All these and other arrangements have already been agreed on by Raaba Gada and the Government authorities, involving even the President. The Kenyan Government, too, has promised support in this matter.

Rooba Guyo (another Kenyan elder):
If I may briefly say something on the negative impact of the campaign to return the OLF youths, I would like to point out that when the OLF first came to our region, we welcomed them and asked them, 'who sent you here?' They said, 'we have joined the OLF on behalf of Raaba Gada and in defence of the Oromo nation'. We then gave them shelter and food and support. Later, after they were established, we heard that the Abba Gada had made a call to them to return home. Because we had to obey the decision of Raaba Gada, we held a series of meetings to discuss among ourselves on how to deliver this message to the OLF who were by now established in remote villages away from our reach. Our campaign to implement the call by the Abba Gada was met with strong opposition from among ourselves and the OLF leaders. We held peace meetings even with Ethiopian communities to declare total peace in the area. Some of us who led the peace initiative were threatened with murder. If initially the OLF had the main objective to form a resistance against the EPRDF and to be the armed force of the Raaba Gada, why then are they refusing to heed the call by the Raaba Gada to return home? There must be a hidden agenda which we ought to unearth. Some major clans and even individuals have taken advantage and have re-deployed OLF forces to suit their own unstated interests. Some of these youths have been trained as terrorists to carry out clandestine murders of fellow Boran and others in the area. We know very well that Oromo organizations do not promote petty politics and local intertribal conflicts like the one we have at our regional level. The Oromo aim is to unite all Oromo and even the neighbouring Somali communities for a common cause. Even those inclined towards Somali politics. The majority of us who support the Raaba Gada resolutions would want the quick return of the OLF so that peace prevails in this area. Those others who have expressed that they can only make the decision if the Member of Parliament of the area has been involved, should not dismiss the resolution of the Raaba Gada now but send a word

to the MP to come and give us his mind. It is our suggestion that the MP should personally discuss with the Raaba Gada council.

The meeting thereupon unanimously agreed that the Area MP who represents the views of the Kenyan Boran and the Government should host the next meeting with the Abba Gada when he is ready.

The *hayyic*[5] Diima Diida continued to explain that major problems resulting from Boran support for the OLF included:

— Government failure to give Boran recognition in both countries.
— All development activities in Boran region stalled.
— Military presence increased in the region.
— Livestock of families supporting the OLF was seized in Ethiopia, even if it belonged to Kenyans.
— Parents and relatives of those youths who joined the OLF were jailed without trial in Ethiopia.
— Ethiopian Government interference with movements of people from village to village and from one zone to the other and from Ethiopia to Kenya, affecting even intermarriage.
— Ethiopian Boran livestock traders were held back from marketing their livestock in Kenya.
— Even ordinary Ethiopian Boran who wanted to sell one or two heads livestock for household needs were not allowed to sell them in Kenya because the Ethiopian authorities suspected them of taking the animals to the OLF.
— All those men and women who were employed in the Ethiopian government service were dismissed if they were suspected of being OLF sympathizers.
— Gabra and Garre youths were recruited into the Ethiopian police force and the army to the disadvantage of the Boran.
— The Gabra and Garre communities in Ethiopia were given a share in power and occupied former Boran positions.
— Relocating parts of traditional Boran land to Region Five of the Somali portion, such as Goof, Laee, D'okkisu and a large portion of Moyale town.

Then the Abba Gada asked the gathering to tell him one or two advantages that Boran gained from supporting the OLF. A Kenyan elder responded:

If someone travels, it is important to know your destination and for what purpose. As the situation is now in Kenya and elsewhere, every ethnic group has an army of their own such that no enemy tribe can easily attack them. Likewise, we Boran, we who have many enemies around us today, get advantages from the presence of the OLF in this region. The Warr Dasse community now fears to attack us and even the government, unlike in the past, recognizes us. That is the only advantage OLF has for us.

However, because of our support for them, many of our leaders were murdered. Some who withdrew that support were murdered by the OLF themselves. They have lived among us and had affairs with our wives.

---

[5] *Hayyic* is a special singular of *hayyu* which puts the emphasis not so much on the institution or the office but on the individual person.

They have provided security to our *fora*[6] livestock and as well on the main roads where the bandits used to rob vehicles. The roads are safe now. However, if I come back to my first statement, what does the OLF aim to achieve, and why did they establish themselves in this region, a country which even does not belong to them? If we enquire from those Boran who have travelled into the interior of Ethiopia we get the following information:

— That the president of the country is an Oromo.
— That all local administrators of the locations, districts and even at provincial offices are Boran, our own children and hence Oromo.
— That all security officers in this province are Boran youths, hence Oromo.
— That the official language of Ethiopia[7] today is Oromo for official communication in schools and offices.
— That the flag of the Oromo is hoisted along with the country's national flag.
— That even in Addis Ababa a majority of officers in central government offices are Oromo.
— That there is nowhere in Ethiopia where there is a build-up of an OLF army. Such a build-up takes place only in this part, an area which even does not belong to Ethiopia.
— That the Boran are the only Oromo community which is opposing the government.

If this is the case, I now wonder what else these OLF organizations want if all that has been achieved. Do they want a separate nation for the Boran community? Where are they leading us to? Where will be the end of their journey? Who else apart from the Abba Gada can tell us what is possible or not possible for the Boran? We do not expect that from the Member of Parliament who only represents the Kenyan Boran of Moyale District and is not a national leader like the Abba Gada. Today we must all agree to get rid of the OLF in this part of our territory, and to do this we must ask our government to use all means to return them to Ethiopia where they came from. Many among us talk of either Tigre bribes or OLF bribes. They are not contesting for anything. There is nothing special in the Ethiopian government's role in Boran territory, just like any other part of the country they have to keep it as part of the nation. The Tigre we always talk of are here just as representatives of the government which, as you have already heard, is manned by both Oromo leaders and them. Many of our people were killed. Why should we, the Kenyans, die for a country which does not belong to us? Where is the common enemy or which other reason is there to bind us together and for which we may die fighting? And why should even our brothers from Ethiopia and anybody else die?

Some of the murders of our people were blamed on the Tigre. If I may ask, do the Tigre know Taaro, or Jaatani Ali? For sure one of us must have led the murderers to the victims. The Tigre do not know the OLF hideout at Kuyaal or in Butiye village. It was people like you and me who led them and showed them the places. So what are the Tigre to be blamed for alone? Others may even have taken advantage of the situation, and that we know if we are to speak the truth. The Tigre do not know the livestock of the youths who joined the OLF, it was you who showed them

---

[6] This refers to the satellite herds which are not maintained at the settlement.

[7] In fact the Oromo language enjoys such a status not in the whole country, but in Oromia, the largest of the federal states of Ethiopia.

that cattle to seize. The Tigre did not know Abba Saara who supported the OLF, it was one of us who directed a finger at him for his arrest. The Tigre did not know the existence of OLF in Kenya; it was some of us who told them. We need not blame the Gabra or the Garre for spying for the Tigre on Boran. They never did any spying on us; it was you Boran who did that among yourselves. These acts of preying on each other were what forced the Raaba Gada council to come so as to arrest the situation from getting worse. Boran, like the government, also want peace to prevail. In the past we have never heard Boran killing a Boran except by accident, but today Boran are killing one another with their own hands. This is a misfortune for the present generation and as well for those to come. We must now adhere to the resolution of Raaba Gada.

In conclusion the Abba Gada reiterated his call to the OLF to return home. He said:

> If you have volunteered for the defence of all Boran, now come home, we shall sit together and agree on a common strategy to defend ourselves. If you are building up arms against the Ethiopian government, give up now since you cannot achieve anything without our support! We want total peace in this region. We want Kenya and Ethiopia to enjoy peace and stability as sister nations. We Boran are the first-born among the Oromo and we want others to emulate our example as a peace loving community. My blessing is on you who hear and adhere to my call. Your support for me will give me confidence and praise in the eyes of the two governments. No government, neither Kenya nor Ethiopia would fail me – this I know very well. I now request you sons of Boran not to fail me on your part.

Other local leaders made a similar call in support of Abba Gada. The meeting closed with the decision to wait for the arrival of the Area Member of Parliament who was expected to arrive after two weeks.

That next meeting took place half a year later, from 14 June to 16 June 1997. The MP Mohammed Galgallo, the host, was only present on the first day and then left after a row. On the following day he was formally ex-communicated from the Boran fold, cursed, and asked to apologize to the Abba Gada.

People, 4 July 1997 (p. 9) summarizes the turbulent events very well:

> The cursing took place on June 15, a day after Galgalo publicly asked Boru Madda, the spiritual leader (Abba Gadha) who he had invited to the meeting at Moyale, whether he was aware that Borans living in Kenya were being killed by both Kenyan and Ethiopian security forces and whether he had the jurisdiction to bring the killings to an end. Said a source who attended the June 14 and 15 meeting: 'The spiritual leader was angered by Galgalo's questions, reasoning that his authority had been ridiculed.' [. . .]
> At the meeting the following day, Madda 'pardoned all the spies and those collaborating with the rebels from Ethiopia' and asked all to live in harmony. But the issue of Galgalo questioning the spiritual leader arose and it was at this time that the assistant minister was 'cursed' [. . .]

45

> Because of the 'bad omen' that may befall the residents if they take side with Galgalo against the Abba Gadha, the source and a group of youths have appealed to the Moyale MP 'to apologize to the spiritual leader for the sake of the entire community'.

In a written statement entitled 'The Resolutions of the Borana Raba-Gada Peace Conference: 14th June 1997 to 16th June 1997' signed by Hajji Wario Guracha, the Paramount Chief of the Kenyan Boran in the Government structure, and Boru Madha Galma, the Abba Gada (by the latter in Amharic characters), the excommunication is pronounced as follows:

> P/S. The Aba Gada, Boru Madha Galma, wishes to notify both governments that, Mr. Mohammed Galgalo Duba is pronounced by him as not being a Borana. He should not be given any responsible position as a Borana leader by any of the governments. This is a consequence of his effort to undermine the juridical powers of my GADA, and behaving in a disparaging manner during the peace conference [sic] held at Moyale Kenya.

The Muslims in the Boran community tried to encourage Mohammed Galgallo by saying that, as a Muslim, he could not be affected in any way by the curse of a mere pagan. But, as election time approached, Mohammed Galgallo appeared to be tied down by business in Nairobi, and even his supporters started to wonder what was preventing him from campaigning. He did not show up for the KANU (Kenya African National Union) nominations and hence did not contest in the elections.

After the elections a number of seats for nominated MPs were allotted to the various parties. One opposition party, the Luo-dominated FORD (Forum for the Restoration of Democracy, Kenya) nominated him, after he joined that party.

It was thus, in combination with his misjudgement of Boran internal power relations, the OLF issue which led to the loss of Mohammed Galgallo's power base in Moyale.

The OLF was gradually losing its foothold in the area. It did not help them to have a good relationship with the Degodia among whom some of their units lived. The Degodia themselves were being estranged from the Boran and developed an alliance with the Garre instead. They had supported the winning Boran candidate in the December 1997 elections, Dr Gurrach Boru Galgallo, KANU. This led some of the losers to stir up feelings against them. But the main factor leading to this rearrangement of forces was a conflict over arable land.

On the eastern outskirts of Moyale town there is a sub-location called Kinisa with a Gabra (Miigo) sub-chief. Further towards the east are areas inhabited by Garre, including Yaballo sub-location. In 1998 the 'Yaballo farmers' co-operative society' was founded by Boran, Gabra and Burji to claim the fertile land in this area and to prevent Garre from 'encroaching' on it. Garre and Degodia thus found themselves in the same category of 'foreigners' encroaching from the east (Mandera and Wajir Districts respectively) on Boran ancestral lands.

In August 1998 ninety-nine elders, listed by name at the bottom of the document, met in Butiye, a Boran-dominated sub-location of Moyale, and agreed on the following resolution which they asked Shongolo to put into English and type up for them. We here render it in a shortened and slightly edited version:

Land Encroachment – Moyale District:

A meeting was held today 13/8/98 at Butiye to discuss recent land encroachment from the south of the district by the Degodia community. The meeting was attended by over one hundred elders and youths. It comprised representatives from sub-locations such as Butiye, Heilu, Somare, Bori, Sessi, [Moyale] Township, Odda, Godoma and Dabel. The communities represented were mainly pastoral communities such as Boran, Gabra, Ejji [the Boran name of the Isaaq Somali, who were represented by two local traders], Sakuye and Burji.

The basic issue discussed was the district's vulnerability to constant encroachment and exploitation by nomadic communities from North Eastern Province, namely the neighbouring district of Wajir. Over the past century, the indigenous community of Moyale District have lost access to considerable tracts of land along their eastern border and of recent along their southern border. This was as a result of illegitimate expansion of pastoral communities from North Eastern Province who come into the district in times of crises such as drought and ethnic hostilities. Their settlement in this district has led to tremendous over exploitation of pasture and water resources and severe damage on the pasture land. Native communities are presently facing severe pressure from the outsiders not only on pasture lands but also [on arable] land ownership rights. In the past the indigenous communities of Moyale District have approached these challenges of forceful occupation of their land through dialogue and peaceful means. Today the idea of protecting and preserving enough land for both present and future generations is confronted with a new dimension – the claim over land rights. With respect to state policies in force, the indigenous communities of this district cannot on their own defend their traditional rights to their pastoral areas and its resources from incursions and occupation by outsiders. They therefore heavily rely on the state to protect the territorial integrity of our pasture lands and its resources. The meeting was convened in the wake of recent claims over parts of Moyale district by the recent arrivals into the district – the Degodia. The Degodia from Wajir District came into Moyale District in early 1993, after the devastating drought of 1991/92, which drastically dislocated the pastoral communities of the district. The affected pastoralists who lost about 95% of their livestock, moved to town centres to eke for their survival. The Golbo plains were thus left uninhabited for some years. The Degodia who too were greatly affected by the same drought fled into Moyale and other districts and sought refuge among the indigenous communities. Some of their large herds of camels had survived. The Moyale community gave them warm hospitality and allowed them access to pasture areas of the uninhabited Golbo plains. For several years the Degodia have had access to many parts of this district without restrictions. They have lived amicably and in peace with the indigenous communities. In the meantime the local pastoralists, who lost their livestock, have started rebuilding their herds and resettling back to their former settlements. This resulted in a great pressure on the pasturelands. The local community started expressing discontent over the Degodia occupation. When the Degodia were asked to move further south from the overgrazed areas, they showed some resistance and reacted by hostilities. The aftermath of this was seen as a situation that will lead to ethnic conflict which the local communities would not condone.

Recently the Degodia have established a makeshift settlement at Illaadu [some fifteen kilometres from Moyale on the Moyale-Dabel road]. They have made a strong presentation to the provincial administration that they be given equal rights of occupation in this district with the indigenous communities. They have as well requested that Illaadu, their temporary settlement, be given the status of a sub-location, and that a member from their community be appointed an assistant chief to the area. They have also made false claims to parts of this district and that they have lived within the district for over sixty years.

From the foregoing it is clear that the Degodia have sinister motives which will eventually cause ethnic clashes in this district. This is an open aggression to our traditional territory. In order to curtail these mushrooming problems between the indigenous local communities and the Degodia, the meeting unanimously made the following resolutions, suggestions and requests to the state authorities concerned with this part of Kenya.

1. Degodia are not indigenous residents of this district. It is resolved that they should go back to Wajir District with immediate effect.
2. Their allegation that they have farmlands in Moyale District is baseless and the local communities would not allow Degodia settlement in this district.
3. In the recent past, foreign immigrants into this district were issued with ID cards. Even adults who were registered in other districts have been issued with new ID cards bearing the names of locations in Moyale District.
4. The Degodia have established a settlement at Illaadu without the consent of the local community and the leaders. That is to be considered an illegal settlement and the Government should with immediate effect disband it, since it is a risk for security, peace and stability in the area.
5. It was resolved that the present practice of issuing ID cards and changing particulars to resettle more people to this district should cease with immediate effect and that the government should launch further investigation into the cases in which such papers have already been issued.
6. Degodia came to this district as refugees who fled from North Eastern Province during the Degodia/Ajuran war and the subsequent drought. The indigenous community provided them shelter and access to pasture land temporarily while reconciliation was going on back in Wajir District. After peace prevailed, they declined to go back having sinister motives to claim part of this land. It is now resolved that since peace is prevailing in North Eastern Province, we strongly appeal to the government to move the Degodia back to their district before anticipated tribal hostilities flare up in the near future.
7. Past experience has shown that wherever Degodia have settled with other communities, as at Hola, Garissa, Wajir, Mandera and Isiolo, there have been tribal clashes provoked by the Degodia themselves.
8. That we have no confidence in the following elders representing the Degodia: [. . . here two County Councillors, a businessman, a Kadi and a driver for a Ministry are enumerated by name]. The above have

incited the Degodia to claim this land and have thus threatened our peace. The two councillors were leaders in the Degodia/Ajuran war and we have no confidence in them at all. One of the councillors is known to have been a firearms dealer. His wife was arrested at Negelle [Ethiopia] selling firearms.

9. We enforce the resolution of the leaders' meeting of 2nd July, 1998 that the Government should ask the Degodia to go back to their district.

10. Unlike in the past, Chiefs and Assistant Chiefs should not issue land grazing documents to the Degodia in any part of this district.

11. We elders of Gabra, Boran, Burji and Ejji unanimously pass a resolution that we shall guard the peace and we are a peace loving people and we pledge loyalty to the Government.

12. Stability of both Kenya and Ethiopia. We thank our Government for having appointed a Minister, an Assistant Minister and an Ambassador from these communities.

13. We also resolve that whatever we are confronted with, we shall use peaceful means to solve our problems.[8]

The background to the disagreements about the Illaadu settlement of the Degodia is that both Gabra Miigo and Degodia are camel people. Some Gabra Miigo pastoralists who had settled in Kinisa shared a dam which was midway between Kinisa and Illaadu with the Degodia. Another point of irritation was that the District Commissioner (DC) had provided famine relief food to the Degodia only. It was alleged that the Gabra Miigo then committed some clandestine murders of Degodia, one at Funaan Nyaata and two at Ambaalo in July 1998, to frighten their clans-people away.

The 'anticipation' 'that tribal hostilities' might 'flare up in the near future' (see above 7) is, of course, a thinly veiled threat. The repeated emphasis on a good relationship to the Government, the expressions of gratitude and pledges of loyalty were meant to imply that the Degodia Somali are less good Kenyans and aimed at mobilizing Government forces against them.

The English terminology of the above document, although inspired by the language of officials and meant to sound official, also reflects the Boran terms used in the discussion. That people from other parts of Kenya are constantly referred to as 'foreigners' reflects the spirit of the Boran term *orm*. *Orm*, which is the same root as 'Oromo', is used by Boran with reference to people who were once a part of them and then became alienated from them. This term therefore also has the connotation 'traitor'. The Degodia, who had no credible Oromo roots, were also straightforward referred to as *nyaap* – 'strangers' who are not part of the *nagaya Booraana*, the 'Boran peace' and therefore not only strangers but also enemies (another term for 'enemy' is *sidi*). 'Local community' is the translation of *worr lafa*, 'people of the land'.

This document tries to put the land issue in moral terms. Instead of just

---

[8] As the last of four pages of this document are lost, the end of the text has been reconstructed from Shongolo's handwritten notes.

describing two pastoral communities competing for the same resources in the same area, it imputes 'sinister motives' on one side. The motives are said to be 'sinister' because they involve taking something which belongs to others. The legal situation is by no means so clear. A colonial heritage of closed districts and reserved grazing rights here is at variance with the freedom of movement of all Kenyans within their country.

The terminology is informed by international discourses. The category 'indigenous', originally developed for the earlier inhabitants now forming minorities in white settler communities in the Americas and Australia, may be difficult to apply to Africa where the majority of the population are Africans.[9] Still it has a moral appeal: it implies that the groups referred to by this term deserve a special degree of protection. In terms of length of stay in Moyale (if that is implied by the term 'indigenous') some Ajuran may have the best claim to this status. In a longer perspective, can the Oromo, whose presence in the area goes back to the sixteenth century (like that of the Whites in America), be truly called 'indigenous'? This is not to contest anyone's claim to be a citizen of Moyale. It is just to show the absurdities to which the application of this term leads in an African setting.

To trigger a major outbreak of violence in this situation only one more incident was needed. This occurred on the 22 September 1998. A small Mitsubishi Canter lorry belonging to Mohammed Dadacha, a Gabra Miigo retired army officer was attacked near Funaan Nyaata. It was a passenger service vehicle between Moyale and Sololo. The name of the vehicle, painted on it, was 'Survivor'.

Shongolo's sister's daughter Nasibo, nine years old, was the only survivor of the attack, and from her Shongolo gathered the following details: first a bullet wounded the driver, so he stopped the lorry. Then people came, six men in uniforms, one in a *kikoi*[10] and one in ordinary trousers. The passengers thought they were ordinary robbers and started to pull out their money. 'We do not want money, lie down!' Shongolo's sister lay down on her daughter Nasibo to protect her. The man in the *kikoi* then gave an order to shoot.

With the exception of Nasibo all passengers and the owner/driver were killed. The dead included a brother to the MP and Assistant Minister Dr Gurach Galgallo.

The languages Nasibo heard during the attack were Swahili and Somali. Nothing else is known about the attackers but that they spoke these languages and were not robbers: when later the police came, some of the nine dead were still holding their money in their hands.

The attackers were suspected to be Degodia. This suspicion is based on the following circumstances: the Officer in Command of the Odha military camp at the time, a Major, was a Degodia. The evening before the shooting

---

[9] For a fuller treatment of the Africa problem in the indigenous rights discourse see Niezen (2003) and Chapter 3, 'Feedbacks and Cross-Fertilizations'.

[10] A wrap, common in a number of African countries, similar to a sarong (also *kikoy*(i)).

he dispatched two lorries. They were supposed to go to Marsabit. But they took some food to Funaan Nyaata, where there is a Degodia camp. Mohammed Dadacha, coming with his Survivor from Sololo, saw them unloading. He stopped and asked what was going on. He then reported to the police.

When later people, including the Kenyan Police Reserve, wanted to follow the footprints leading away from the Survivor, they were prevented from doing so by the army who closed the whole area. They then became suspicious of some army involvement in the incident.

The Degodia Major was transferred the next day. During a huge meeting massive complaints were uttered against him to Adam Biru, the Army general who had arrived from Nairobi. Other elders had complained about him in writing to the DC.

The DC himself, who was held responsible by the Gabra and Boran for allowing the Degodia settlements, was transferred a little later. There had been many complaints about his lack of support for the investigations into the Funaan Nyaata attack.[11]

The next month, on the 24 October 1998, it was the Degodias' turn to be massacred, but on a much larger scale.

A huge force of Boran crossed into Wajir District and in the Bodod, Tullu and Muddam area, where there was plenty of water from the El Niño rains. In the morning they called the Degodia, camping there in many hamlets, for a meeting. Also large herds had gathered at one of the ponds, and the herdsmen were called to the meeting. When large numbers of Degodia had gathered, the Boran opened fire on them. In the press this massacre came to be known as the Bagalla massacre, a name which is reminiscent of the Wagalla massacre of 1984. 'Bagalla', however, is not known as the name of a place to Boran.

The Boran then drove away huge herds to the north, presumably into Ethiopia.

Accounts vary as to the numbers of dead. After a week of press reports based on exaggerations and speculations, the *Weekly Review*, 6 November 1998, gave the following breakdown: 'By early this week, the actual number of dead was put at 187 (87 men, 51 women and 50 children), while 83 people were still unaccounted for and 36 seriously wounded, some of them critically.' The magazine goes on to explain:

> The raiders were also reported to have driven away at least 3,000 head of livestock, including camels, cattle and goats. Some reports talk of as many as 15,000 head of livestock being stolen and the use of poisonous chemical bombs in the attack. Such reports have not been verified, and a government attempt early this week to exhume the bodies of the dead for examination and determination of the exact number was called off on religious considerations. The majority of the inhabitants [in fact all Degodia] are Muslims, for whom it is taboo to exhume corpses [. . .]
> Following the killings, the government announced that it had beefed up security in the entire region, with police and military units being moved

[11] Shongolo has given a similar account to Cynthia Salvadori.

to various areas that are likely to be volatile. The government of Ethiopia released a statement soon after the Bagala massacre, condemning the act and absolving its armed forces from any involvement in the incident. Some political and religious leaders from the region have called for a judicial commission or a parliamentary select committee to probe the incident and the establishment of a 'national disaster fund' for victims of the massacre. While it is still not exactly clear who was responsible for the heinous act – which has been blamed by several political leaders from the region on the Oromo Liberation Front (OLF) from Ethiopia that is fighting a secessionist war against the government in that country – the Bagala massacre has caused an unlikely public row between the provincial security teams in Eastern and North Eastern provinces, with the latter blaming their counterparts for the incident.

The *Daily Nation*, Friday, 6 November 1998, also speculated on OLF involvement:

> The involvement of the OLF militia-men in the attack raised lots of questions from Wajir leaders.
> The leaders who claimed that the militia-men had camps at the Kenyan towns of Ambalo, Uraan, Kiniisa and Makutano within Moyale district questioned Kenya's sovereignty.
> Led by the Social Democratic Party national organising secretary Mr Abdi Sheikh Bahalow and the Wajir branch Kanu chairman Haji Omar, the leaders said the armed militia-men who had acted as mercenaries for the Boranas and the Gabras, had the camps in those areas with the full knowledge of the government.

As can be seen from the earlier part of this chapter, direct OLF involvement in this massacre is most unlikely. The OLF camped among the Degodia in Moyale District. Their good relationship with the Degodia had not been affected by the declining relationship between their local fellow Oromo, the Boran, and the Degodia. And anyhow the raiders were suspected to have moved their loot into Ethiopia by the very same media who suspected them of being OLF members. How could OLF fighters have sought refuge with thousands of animals on the territory controlled by a government with which they are at war?

In fact it is common knowledge among Boran, that the attack had been carried out by Ethiopian Boran who, by that time, in their majority were opposed to the OLF. People mostly keep silent about it. One of the rare public occasions on which this was stated openly was a meeting of Ethiopian Boran in Moyale in March 1999, to which the Boran from Moyale, Kenya, had been invited. Twenty-seven people had been killed by a landmine believed to have been planted by the OLF. An elder accused the Kenyan Boran of not having expressed their condolences and showing lack of *borantiti*, Boran-ness. He contrasted that with the helpfulness of the Ethiopian Boran towards their Kenyan brethren. 'The other day, when nine of you were killed at Funaan Nyaata, we immediately sent a force against the Degodia and killed so many of them before you had even asked us to do so.'

The Kenyan public and the Kenyan media, dominated by people who never even think of the vast arid stretches of the north of their country, let alone go there, are, of course, easily misled on such issues. Some of the press reports clearly show the cultural distance of central Kenyans from the north of their country and the extent of their ignorance about it. Again the *Daily Nation*, Friday, 6 November 1998:

> The civilised world, hundreds of miles away was only to awaken to the grim reality of death four days later.

To the *Daily Nation*, it seems to be beyond question that northern Kenya is not part of the civilized world, while central Kenya, irrespective of the amount of violence and corruption which is found there, is a part of it. An age old dichotomy of stereotypes is repeated here: barbarism versus civilization, the desert and the sown.

Even the simplest geographical proportions are distorted:

> By the time the government was informed and security forces were dispatched, five days had passed. The raiders had all the time to cover more than 500 miles towards the Kenyan-Ethiopian border.

The distance to the Ethiopian boundary was, in fact, about 50 miles only (80 kilometres). On the title page of that issue of the *Daily Nation* one finds the picture of a Maasai as an illustration for 'Oromo raiders'. Apparently all nomads are the same.

At the root of this ignorance we find lack of interest. During the parliamentary debate about the massacre many seats remained empty. The Foreign Affairs Minister, Dr Bonaya Godana, complained: 'If 10 people had been killed and they were not from northern Kenya, this House would not have been half empty [. . .]. A while ago a member exclaimed *wacha mali-zane* [let them finish off each other]' (*Daily Nation*, Friday, 13 November 1998).

This level of ignorance around the centres of power in Kenya made it easy for local politicians in the North to play their own identity games.

The former MP for Wajir West[12] in a 180 degrees turn from the earlier situation – the *shifta* war of the 1960s, when the Degodia were Somali irredentists and the Boran Kenyan loyalists – tried to depict the Degodia as good Kenyans and the Boran as separatists. Again the *Weekly Review*, 6 November 1998:

> Khalif and the other Degodia political leaders wonder how heavily armed groups of Oromo tribesmen from Ethiopia, estimated to have been 600-strong, could have crossed unnoticed and unchecked to slaughter innocent Degodia villagers. They blame the new commander of the Kenyan Army, Lt. Gen. Abdullahi S. Adan, a Boran and the provincial administration in the region for neglecting the security of the Degodia,

---

[12] The MP for Wajir West, Mr Ahmed Khalif was re-elected in 2002, but died in a plane crash in Busia when he accompanied the Vice President Kijana Wamalura to celebrate the NARC (National Rainbow Coalition) 2002 election victory.

even when there were abundant signs of tension between the two ethnic communities. [. . .] They talk of an external factor in which the Oromo tribes, through the OLF, are seeking to establish their own separate country to be known as 'Oromia Republic' carved out of northern Kenya and southern Ethiopia. The former Wajir West MP notes, however, that some of the Degodia politicians are equally to blame for the latest massacre through their belligerent utterances, which have tended to paint the clan as expansionist.

A different view was in the same period circulated in the form of a leaflet in Eastleigh, the 'cheapside' of Nairobi which is the bridgehead for all Northerners in the capital and has also a heavy Somali component, consisting both of Kenyan Somali and refugees:

Press release:
'Fear and Tension in Northern Kenya as Residents Arm Themselves'
Investigation conducted by HAKI team reveals that the massacre which occurred in Wajir on 24th October, 1998 where more than 140 innocent Kenyan people were killed in cold blood is connected to the regional politics and especially to the problem in the Horn of Africa. Ethiopia is at war with Eritrea. It cannot therefore stand war on two fronts. In a bid to divert the attention of the OLF to [sic] self-defence, it had to create tension around the border. The Ethiopian Government has under its control members of Oromo Community who are loyalists to its cause. These loyalists are armed and act as police reservists popularly known as TABAQA. Our investigation has revealed that it was this Taqaba [sic] that crossed into Kenya with the blessings of the Ethiopian government and attacked Degodias. Without ascertaining the truth, Degodias blamed it on OLF who in any case are not even based in Kenya as alleged.[13] The truth is that the entire NORTHERN KENYA is now griped with tension. The stage is now set for full scale war as Degodias raise more than 27 million Shillings to purchase arms. Reliable sources confirm that 5 Degodia tycoons led by Sheikh Ibrahim Burhan lead this operation. 3 other Ibrahims and a major retired Mohamed Unshur assist in the operation. Mr. Unshur owns a plane that transports miraa[14] from Nairobi to Mogadishu and has been instructed to bring arms on his way back. Moreover the Degodia have formed a hit squad to assassinate prominent Boran personalities in Nairobi. Top on the list are:
1.   Hon. Dr. Guracha Galgalo [The MP Moyale]
2.   Hon. Dr. Bonaya Godana [The Foreign Minister, MP Marsabit North, in fact a Gabra (Gaar)]
3.   Lt. Gen. Adan Abdullahi [Commander of the Armed Forces]
Leading the operation is inspector Issa Mohammed of the notorious flying squad assisted by one Diba Gayo a BORAN who has been commissioned to monitor the movement of those on the hit list. On the other hand, BORAN in Kenya have appealed for reinforcement from their counterparts in Ethiopia. Impeccable sources reveal that over 1000 armed people are being held at IDDI LOLLA[15] in Ethiopia awaiting a signal from Kenyan Boran in case of any attack. These and other scheming going on in Northern Kenya if unabated will lead to a terrible incident of ethnic

---

[13] In fact, the OLF was still present in Kenya in this period. They were only expelled in February 1999, three months later.

[14] Sw. *miraa*, Arab. *qat*, Amharic *chat* is a mild stimulant in the form of the green leaves of *catha edulis* which are chewed.

[15] Opposite Sololo on the Ethiopian side of the border.

cleansing unheard of in the history of Kenya. We appeal to the Government, local leaders both political and religious leaders and Wananchi[16] in general to do all things possible to avert these crises.

Parts of this document, like the cryptic allusion to the '3 other Ibrahims' whom apparently the authors do not dare to name fully, smack of rumour mongering. Other parts, like the reference to the war between Ethiopia and Eritrea as having repercussions in the wider area, need to be taken seriously. All the three people on this list much later, on 10 April 2006, died in a plane crash. This crash has, however, not been linked to a Degodia plot.

Although OLF involvement in the Bagalla massacre is highly improbable, this event nevertheless was one of the steps which led to the military defeat of the OLF at the hands of Kenyan forces, and their ultimate expulsion half a year later. This becomes clear from the report presented by Marsden Madoka, Minister in the Office of the President, in June 1999. 'The minister said the investigation team could not rule out completely the involvement of the OLF militia in the massacre.' As no other culprits were at hand, all attention was directed at the OLF. 'Madoka said the issue of the OLF needs to be addressed as it has the potential of disrupting peace in the region, adding that trafficking of illegal firearms along the border was rampant and a matter of grave concern.' (In fact the OLF are not known for selling their arms.) In the same press conference Madoka also blamed Dr Gurach Galgallo, MP Moyale, 'for making inflammatory statements that led to the massacre' (*Daily Nation*, 23 June 1999). The accused MP and Assistant Minister called the report a bogus report. There has been feuding in the area long before the OLF presence (*Daily Nation*, 24 June 1999).

The period between the Bagalla massacre and the expulsion of the OLF was characterized by a shift in the ethnic composition of the nomadic population of Moyale District. Guyyo Galgallo, a Boran elder,[17] explains how the Ajuran took advantage of the dissent between Boran and Degodia:

> The Ajuran played a 'monkey role' in the aftermath of the Bagalla massacre [Monkey is the trickster in some African folk tales]. They were peddling information between the Boran and the terrified Degodia. Some day they send reports to the Degodia saying that Boran forces were on the way coming for them again. The Degodia fled beyond Wajir in fear of the attacks. At the same time, the Ajuran also sent reports to the Boran settlements in Ambaalo and Bori saying that a huge Degodia force with sophisticated weapons from Somalia was advancing towards the Boran settlements to avenge the Bagalla massacre. The Boran settlements withdrew back to the Sololo area although pasture and water were scarce there.
> The Ajuran thus succeeded to remove both the Degodia and the Boran from a large area where there was good pasture and plenty of water for the livestock. The Ajuran then settled in the area with ease.
> Prior to this the Ajuran had attempted to obtain these grazing rights in vain. They had been sending delegates to the Boran local leaders,

---

[16] *Wananchi* is Swahili for 'children of the soil' or 'citizens' (of African descent).

[17] Interview of Guyyo Galgallo with Abdullahi Shongolo, June 1999, Butiye, Kenya.

requesting permission to settle in the area from where the Degodia had fled. The Boran only promised to discuss the matter among themselves later.

The reason why the Ajuran wanted to occupy the Bododa area was that it had plenty of water trapped in the *bule* rocks [i.e. in the lava fields] from the El Niño rains[18] as well as plenty of pasture for all kinds of livestock.

Most of the Ajuran *fora* livestock were grazing in Batalu and areas bordering the Garre territory to the north of Wajir. There is an on-going conflict between the Garre and the Ajuran over the border of the two Districts of Wajir and Mandera. In search of pasture the Ajuran were penetrating into Mandera District which was already under pressure from Garre livestock. While this conflict was escalating, many Ajuran withdrew to the Bododa.

The Ajuran, however, were not satisfied with the Bododa area. After about two months or so they finished the water at Bododa and moved to Funaan Nyaata dam, where the Boran and the Gabra were already settled. The Ajuran elders on several occasions sent delegates requesting even for water at Holale [close to Moyale town]. Realizing that the Ajuran like our old step brothers [an ironical reference to the Degodia] are invading our territory by means of 'cold war' [using the English expression], we met to discuss about this recently. Some local leaders met with the District Commissioner to discuss the Ajuran incursion. The local leaders representing the Boran, Gabra, Sakuye and Burji were led by the District County council Chairman Gollic Galgallo and Haji Wario Guracha, the paramount chief. The Garre councillors did not attend the meeting.

According to Gollic Galgallo, the District Commissioner Moyale had already made a request to the District Commissioner Wajir, to arrange for the return of the Ajuran nomads to Wajir District. We told the DC that unlike the former DC who helped the Degodia to settle in the area, he himself should not take this matter lightly. We cautioned the DC that should any conflict arise as a result of Ajuran presence in the area, he should be held responsible. The DC agreed to take immediate action to return the Ajuran.

Despite our effort to avoid conflicts over grazing areas and the scarce water, some Boran elders seem to support the Ajuran plight, because like the Boran and Gabra they too are enemies to the Degodia. The first wave of the Ajuran who came were the Gelbaris. They reminded the Boran elders of their past relationship during and before the colonial period, and also of the time of the '*Worr Libin* confederation'. Unlike other Ajuran clans, the Gelbaris speak fluent Boran and are familiar with the Boran *aada/seera*[19]. On their arrival in the area, they came to the homes of Gona[20] elders and requested for re-admission into the Boran fold, the status they had lost many years ago, when they forced the Boran out of their territory. They blamed both the Garre and the Degodia for having taken over the position closer to the Boran which they claimed as theirs. An Ajuran elder claimed, 'we were brothers even long ago and we shared common customs with you unlike the Garre and the Degodia who were both inclined towards the Somali ways in those days. For all these reasons we now ask the Boran to fully accept our return'. We listened to their endless talk but only asked them to return to their district.

---

[18] The exceptional rainfalls of 1998 which caused disastrous floods in parts of the Horn were attributed to the El Niño weather phenomenon.

[19] Customary law; both terms are of Semitic derivation.

[20] Gona is the Boran moiety to which the Ajuran were affiliated at the time of the *Worr Libin* alliance.

The OLF, which had withdrawn into Moyale District in 1991, for the first time launched an attack from this area into Ethiopia on 16 January 1999. According to local sources the OLF attacked an Ethiopian military post at Idilola and another one at Tuqa, just opposite Sololo. They also planted landmines on the roads leading to these places, presumably to inflict damage on other forces which they expected to rush to the rescue of these two posts from Boku Lubooma, a bigger military base. The Ethiopians under attack killed four OLF fighters and then pursued the remainder onto Kenyan territory. The OLF, who carried some injured, went to Dambala Fachana, the nearest village on the Kenyan side of the border. There the Ethiopian forces who followed them got involved in an exchange of fire also with local 'homeguards' (Police Reserve). The same night some of the landmines which had been planted by the OLF on the Ethiopian side blew up and killed about twenty people, among them some of the wounded who had been taken out of the attacked locations by civilian small lorries (Canter) used for passenger traffic.

The Chief of Dambala Fachana reported an 'Ethiopian attack'. The police came and took the injured OLF fighters and two homeguards to the Sololo Mission Hospital, apparently assuming them all to be Kenyans. The same day an Ethiopian military commander complained to the DC Moyale that Kenyan homeguards and Kenyan-based OLF had attacked Ethiopia. The DC denied the presence of OLF on Kenyan soil and reciprocated by blaming the Ethiopians for a military attack against Kenya. He then went to Sololo on a fact finding mission and met the wounded. Without further ado they told him that they were OLF fighters who had launched an attack into Ethiopia. The DC thus realized that he had been misled. He immediately took the wounded to Moyale Hospital. The following day they were taken to court for being illegally in Kenya and sentenced to prison. While still in hospital to recover sufficiently to start their prison terms, the fighters were also visited by the DC and his Ethiopian counterpart, accompanied by Ethiopian intelligence officers who knew all the fighters personally. The same night there was an attack against the hospital, which is right on the border. It was beaten back by Kenyan forces. It is believed that the Ethiopians had tried to abduct the prisoners. The same night these latter were taken to Marsabit and later on to Meru.

Cross-border incursions and unresolved murder cases were frequent in this period. Under the headline 'Leaders Tell of Fear', the *Sunday Nation*, 4 February 1999, explains:

> Top politicians from Isiolo, Marsabit and Moyale Districts have expressed fear for their lives.
> This follows accusations by the Ethiopian government that they were supporting activities of an Ethiopian rebel group.
> Assistant Minister Charfano Guyo Mokku told journalists in Isiolo town that they had informed the provincial security committee of their fears during a recent meeting chaired by the Eastern P.C. [Provincial Commissioner of Eastern Province], Mr Nicholas Mberia, in Embu.

Mr Mokku said politicians linked to the activities of the Oromo Liberation Front rebel group include three assistant ministers from Moyale, Marsabit [sic], a Minister and a senior army officer from the Boran community. 'We have received reliable information that our names have been mentioned by the Ethiopian government and linked to the Oromo rebel group fighting to remove it from power,' said Mr Mokku.

The Minister said the information [. . .] will not be taken lightly as the Ethiopians have in the past made incursions into Kenyan territory to pursue alleged rebel sympathizers and even killed a religious leader [the Imam Haji Hassan] in Moyale town last month.

He said the names of Kenyan Borana politicians should not be dragged into the internal politics of Ethiopia.

'We serve a sovereign state [. . .] and [. . .] should be left in peace.'

Earlier, the Minister had told a security meeting at the Isiolo Baraza Park that the government should not allow the killing of innocent Kenyans by Ethiopian soldiers who cross the border on the pretext of pursuing rebels. He said the Ethiopian Tigre-led government should resolve its power struggle without involving Kenyans in the three districts.

In August last year, Eastern Deputy PC David Chelongoi gave members of the Oromo rebel group an ultimatum to leave Kenya. However, the government has denied the group's presence in the country. [. . .]

The murder which probably shook Moyale residents most was that of the Imam of their mosque, Haji Hassan.

On the night he was killed, armed men believed to be Ethiopians opened fire around Somare Village overlooking Moyale town. It is instructive that nobody in the *manyatta*[21] was killed, giving credence to the view that the shooting was diversionary. The Imam was killed in his Moyale Town residence as shooting raged at the *manyatta*[22]. (*Sunday Nation*, 14 February 1999)[23]

The same day the two District heads, the Kenyan and the Ethiopian one, accompanied by security and other officers and many elders met at the barrier. The Kenyan DC asked for the return of two Kenyan government vehicles which had been confiscated in Ethiopia in response to stone throwing attacks on Ethiopian vehicles on the Kenyan side which were a response to the murder of the Imam. The conversation was in English and went roughly like this (as there were several people voicing the Ethiopian position and as many Kenyans responding to them, it is difficult to reconstruct who said what. We therefore lump the interlocutors together as the Ethiopians (E) and the Kenyans (K)):

[21] Swahili ex-Maa(sai) for a nomadic or semi-permanent settlement.

[22] Italics in the original.

[23] The quote stems from a special report entitled: 'Moyale: Town with a war on its doorsteps', which is also very instructive in many other ways. It depicts a central Kenyan journalist (Gahika Weru)'s impression of that town in the following words: 'Moyale Town, on the border of Kenya and Ethiopia, is Kenya's last frontier in every sense – inaccessible, poor, almost alien and tough to live in.' In spite of all this, of course, for thousands of local people it is the centre of their world. It also gives pictorial illustrations which contrast the Kenyan part of the town with the Ethiopian one, 'a modern urban area with tarmac roads and other amenities [. . .]'.

E: 'Why did your people attack our cars?'
K: 'Because you killed our Imam.'
E: 'How can you claim that? Just a few hours ago at the burial even your DC admitted that it was not known who killed him.'
K: 'Your security people were seen to withdraw into Ethiopia after the incident.'
E: 'Is that enough evidence for you to attack our vehicles?'
K: 'Just give us back our government vehicles. Anyhow it is known that you constantly kill our people.'
E: 'It is not our people who murdered the Imam. And even if it was us, was he not one of us as well? Do you know where he is from? Maybe the only mistake was that he was killed on your soil.'
K: 'How can you claim that he is yours? He is our well-known Imam and the leader of the Muslim community in this district.'
E: 'Do you know his tribe?'
K: 'Well, Boran or Somali or something, he was just a Kenyan from Moyale.'
E: 'So you do not even know. He was Arsi. Is there an Arsi community in Kenya? So he was ours!'
K: 'He was known even to the President of Kenya as the Imam of this town. And even if he was Ethiopian by birth, would that have given you the right to kill him inside Kenya?'

At the end, the two delegations separated in anger. What is instructive about this dialogue was that the Ethiopian officers, instead of a notion of human rights and civil rights, seemed to have a sense of ownership of anyone born in Ethiopia or being of Ethiopian blood. This type of ownership seems to comprise the right to kill.

The Imam may have been suspected by the Ethiopians to help oppositional Arsi to get into Kenyan exile and to collect money for the OLF. Similar suspicions might have been sustained by the Kenyan police who had searched his house some days before his death.[24]

Subsequent to this some prominent elders were either killed or attacked in Sololo area. For several years they had been blamed as supporters of the OLF by Kenyan and Ethiopian authorities, and it is alleged that they withdrew their support due to pressure both from the Kenyan and Ethiopian Governments.

Molu Billida, a prominent elder was murdered in his house on 15 March 1999. Two days later, a well-known, retired senior chief of Sololo, Huqqa Guleid, was attacked in his house but escaped unhurt. Both of these attacks were blamed on OLF by the residents of the area.

Some headlines from this period include: 'Kenyan Security Men Kill Three Ethiopians' when 50 armed men in uniforms from Ethiopia were reported to have attacked Waiyegoda village near Sololo (*Daily Nation*, 11 March 1999), 'Moyale Tense as Families Flee Fearing Another Raid', when 4,500 people were camped at a church compound in Sololo (*East African Standard*, 12 March 1999), 'Addis to Blame for Raids; Rebels Claim', when the OLF and another group, the ONLF (Ogadeen National Liberation Front) from the Somali Region (Region 5) of Ethiopia issued a joint fax

---

[24] *Sunday Nation*, 14 February 1999.

statement 'that Ethiopia is sending assassins squads to sovereign neighbouring countries including Kenya and Somalia to eliminate members of opposition organizations. [. . .] It is the killers who boastfully narrate how they entered Kenya's territory under the pretext of pursuing rebels and killed Kenyans.' (*East African Standard*, 15 March 1999)

In response to numerous violent incidents, there were 'Calls to Expel [the Ethiopian] Envoy amid Hostility in Northern Kenya', as the Marsabit correspondent of the *Daily Nation*, Said Wabera, reports on 18 March 1999:

> Led by the MP for Moyale, Dr Guracha Galgallo, the leaders [. . .] urged the residents to arm themselves since the government had failed to protect them from external aggression. [. . .] The deputy Parliamentary Chief Whip, Mr Mohamed Shiddiye, blamed the escalating insecurity on laxity by the provincial administration, saying that the area was now an 'AK-47 region' [i.e. a Kalashnikov land].

On 22 March 1999, a 'Compensation and Confession Day' was held at Moyale, Ethiopia. Top government officials led by Lt. General Abba Duula[25] and Ato[26] Gabre and two top 'traditional leaders', Jillo Boru representing the Abba Gada of the Boran and the Guji Abba Gada T'ant'allo attended a function where victims of landmines at Tuqa and Iddilola were compensated. The owners of the three vehicles which were destroyed by the landmines were each given a new five-tonne Isuzu lorry. The wounded were given three thousand Ethiopian Birr and for each of those killed five thousand Ethiopian Birr were paid as compensation. The presentations were conducted by the two *Gada* leaders, at a ceremony attended by people from both Kenya and Ethiopia. This earned the government praise as it had shown concern for the people of the area. Jilo Boru expressed gratitude for the government's commitment to care for those attacked without any reason and especially to the General Abba Duula and Ato Gabre for their support of the peace initiative in the region. Later at a separate function, about fifteen OLF members who were either taken prisoner on the battlefield or arrested in town were brought to narrate their activities. They admitted participation in the recent attacks.

A meeting was held on 30 March 1999 at a hotel in the Ethiopian part of Moiale. Ethiopian officials had been invited to discuss recent conflicts. The meeting was also attended by the Security Committee of Moyale, Kenya, and local leaders. A Kenyan Councillor from Sololo was refused participation in the meeting and was forced to return to Moyale. He was so frightened that he asked the Kenyan DC for his Land Rover and an escort. He had been accused of being an OLF supporter; his denial had not been accepted. The head of the Ethiopian delegation, Ato Gabre Allemseged accused also other Boran local leaders of failing to take any initiative to expel the OLF from Moyale District. Supporting this accusation, Mega District Head, a Boran, said:

[25] This name is a descriptive term for a general. *Abba duula* means 'father of war' and is the title of a Boran war leader. The following paragraphs are from Schlee (2003).

[26] *Ato* is Amharic for 'Mr'.

You have failed to support us on many occasions to remove the OLF from among yourselves, while we have given you all the support you needed, some even without you asking for it. Recently, many Boran were killed by people who came from your homes. One day a Boran is killed on this side and the following day one on your side. Nobody else is being killed. The murderers from the two sides are no doubt just Boran themselves. Whenever someone is killed or a village attacked in Kenya, you just allege that the killers are Tigre. There are no Tigre here in Moiale within the military force. The only Tigre in Moiale is this one here, Ato Gabre, whom you all know very well from his role as a peace seeker. Certainly none of you would accuse him of participating in the killings. He has always been advocating peace both in Kenya and Ethiopia. He has done a lot for the Boran including you Kenyans. You know very well that had it not been for him, today the Boran would have lost their land and everything else. He has gone beyond any limit to support the Boran against many odds. He has been accused by Garre of giving too many privileges to the Boran. That is what the Garre complained about to the central government at Addis. We all know that it was all his effort that brought peace, stability and unity among the many communities in this region, and as well between the two sister nations of ours. (from field notes by Abdullahi Shongolo)

The Moyale District Commissioner, leading the Kenyan delegates, expressed similar sentiments and wondered why only Boran were being killed day by day. In concluding, Ato Gabre advised the Boran local leaders in strong terms to support the two governments' effort to remove the OLF from the region. The local leaders blamed the Kenyan District Security team for showing laxity in taking action against the OLF although they, the Boran leaders, had long ago given their consent to such action. The leaders promised to launch a massive campaign to see that the OLF are expelled from the area.

In response to the local leaders' accusation, one month later the District Commissioner interdicted a Chief, who had earlier misled him on the clashes at his location, and the Assistant Chief of Bori sub-location for alleged support of the OLF in their area of jurisdiction.

On 10 May 1999, it was reported that a senior public works officer was killed when a government Land Rover was blown up by a landmine in the Konkona area, 25 kilometres east of Moyale.[27] On 26 May, one could read that 'A joint military operation to flush out bandits involved in laying land-mines on Moyale/Marsabit highway has begun. [. . .] patrols on foot, involving anti-mine armoured vehicles and army aircraft, will cover Dabel, Somare, Sololo, Walda, Uran and Golole in Moyale District'.[28] The 'bandits' struck at were OLF fighters, but it was by no means clear that the OLF had actually laid the mines. It is difficult to imagine why they should have done so, given their already precarious status in Kenya. Rumours that the mines were laid by people from the Ethiopian government side to provoke a Kenyan crack-down on the OLF (who were known to possess landmines, because they had used them on the Ethiopian side) have a measure of

[27] *East African Standard*, 10 May 1999.
[28] *Daily Nation*, 26 May 1999.

plausibility. The Moyale MP had remained vague on the issue and blamed the landmines on an unnamed 'armed terrorist group operating from a base in Ethiopia'.[29]

Finally, in June 1999, after their presence there had been denied for years and action been taken up against them only slowly and hesitantly, military pressure on the OLF reached a level which made them leave Kenya and move into Somalia.

> Two Oromo Liberation Front rivals [meaning rebels] were killed and 13 others captured during an operation by security forces in Moyale District.
>
> Moyale DC Stephen Kipkebut said yesterday that the operation has also recovered eight AK-47 [Kalashnikov] rifles, five different types of guns, 4,037 rounds of ammunition, 362 military uniforms and 50 hand grenades. (*Daily Nation*, 9 June 1999)

Of course, the confiscation of these meagre supplies by no means depleted the resources of the OLF. A little later the OLF was reported to have moved down through Somalia to the coast in orderly and well-armed convoys of vehicles, clearing the way before them of the road blocks set up by the numerous Somali petty militias. They were believed to be on their way to the area controlled by a new ally, Hussein Aidiid, from where they would embark by ship to Eritrea, to continue their fight against the Ethiopian government from the north, having been defeated in the south (from BBC World Service broadcast at the time).

Eritrea had gained independence from Ethiopia after thirty years of struggle in 1992. The leaders of the two countries had been comrades in arms against the Mengistu regime which had collapsed in 1991. There was peace between Ethiopia in its reduced shape and the newly formed Eritrea until 1998, when a border conflict flared up between the two countries and quickly escalated to a full-scale war.

The countries of North-east and East Africa have a long tradition of supporting dissident movements against their neighbours. With peace between Ethiopia and Eritrea the options open to players of this game had become somewhat reduced in number, but with the outbreak of renewed hostilities between the two countries the room for regrouping into new configurations had widened again. One could find a new ally in one of these countries if one supported it against its own dissidents or if one gave support to the dissidents of the other.

As Hussein Aidiid supported both the OLF and Al-Itihad al-Islamiyya who fought the Ethiopian government in the Somali region (Region 5) of Ethiopia, he had become a natural ally of Eritrea. Some of the gratitude of the OLF towards Aidiid might have found its expression in the fervour with which they swept the road to the coast of rival Somali militias. This new configuration has not escaped the attention of Kenyan media, as the following article (in spite of the slight distortions it contains) shows:

'Kenya, Ethiopia Blame Third Force'

Addis Ababa, Thursday
Kenya and Ethiopia today accused unnamed countries [which are named three paragraphs later] in the Horn of Africa of trying to undermine their peace and security.
In a statement published in Ethiopian state media, a joint committee of officials responsible for border security between the two neighbours 'expressed concern over certain countries of the sub-region that are sponsoring and encouraging the activities of terrorist groups, including the OLF and Al-Itihad'.[. . .]
Ethiopia and its smaller neighbour, Eritrea, have been involved in a year-long border conflict that has spilled over into other countries in the Horn of Africa. Eritrea has reportedly sent hundreds of OLF fighters into southern Somalia from where they can move into northern Kenya and Ethiopia. [This must be the wrong way round. The OLF had just been driven out of southern Ethiopia and northern Kenya and apparently were on their way to Eritrea.]
Ethiopia accuses Somali warlord Hussein Aideed of aiding Al-Itihad. [. . .]
The joint committee of border security officials met on Tuesday and Wednesday in Moyale to discuss the deteriorating security situation. [In fact, the meeting took place in Nazareth, Ethiopia.]
Kenyan police reported earlier this week that security forces had killed four Ethiopian soldiers along the border. It was later reported that the four killed were OLF fighters.
Meanwhile, reports from Somalia say that Ethiopian forces have been heavily involved in fighting in and around the central town of Baidoa and by Wednesday had arrived in Dinsor, 250 km south of the Ethiopian border.
Ethiopia has denied that any of its troops are in Somalia. (*Daily Nation*, 11 June 1999)

The last three paragraphs have been quoted because the contradictions they contain illustrate the uncertainty characteristic of the situation.

On 15 and 16 June 1999, the Moyale DC had meetings with the local elders in different localities to stress the maintenance of peace and order as the OLF were leaving the area. At a meeting in Butiye he directed his speech to OLF supporters, saying

'if you still long for them, please part with them with a blessing, but never follow them to their destination. You should also advise them to leave behind the children born to them during their stay among you. The women, too, should not follow them'. He asked the residents to report any members of the OLF left behind. (from field notes by Abdullahi Shongolo)

Almost a month later, on 8 July 1999, the *East African Standard* reports:

Last week's ambush on a Kenya Army platoon by Somali militia [who had taken several army lorries, apparently without meeting resistance worthy of the name] at Ammuma is a spill-over of the war between Ethiopia and Eritrea who are supporting different factions in the war-torn country, Minister of State Julius Sunkuli told the House.
Sunkuli also admitted that Somali warlord Mohamed Farah Aideed was in the country [he must have meant his son Hussein, since M.F.A. had died years before of bullet wounds received in Mogadishu], but denied the Government was paying for his expenses adding that he was here like any other Somali national.

[. . .] Sunkuli explained that the Ammuma incident occurred after Ethiopians who were incensed by the alleged support of Farah Aideed of the OLF, Ogaden National Liberation Front (ONLF) and Al-itihad, [. . .] carried out a pre-emptive attack supported by other Somali factions.
'As a result Aideed and his allies, the Somali National Faction (SNF) [the correct name is Somali National Front] of Maj. Gen. Omar Haji Mohammed 'Masale', moved southwards booting out Gen. Mohammed Said Hersi 'Morgan's'[30] faction from Kismayu pushing them to the Kenya/Somali border', said Sunkuli.
The Government, the minister said, quickly dispatched the D-Company platoon of 3 KAR stationed in Garissa to sort a group of 204 refugees escaping the clashes, when 400 militia-men supported by eight technicals [jeeps with machine guns mounted on top] who were pursuing SSD militia of Gen. Morgan, bumped into them.

Some readers of the *East African Standard* could not hide their amusement about the last paragraph:

The minister told the House that Kenya had strongly warned the warlords that if a repeated incident happened the Kenya Army will act decisively and punish them severely.

The existence of the new alliance Aidiid/OLF/Al-Itihad/Eritrea is corroborated by a report in the same newspaper on the following Monday (12 July 1999):

'Ethiopian Oromo Rebels Go Deeper into Somalia'

Hundreds of heavily armed Ethiopian rebels, sponsored by Somali warlord Hussein Mohamed Aidid, forced their way south through several Somali towns, smashing roadblocks manned by opposing forces, witnesses said yesterday.
The forces were believed to be members of the Oromo Liberation Front. [. . .]
Eritrea, which is at war with Ethiopia, is allied with Aidid and is backing the Ethiopian rebels.
The war between the two Horn of Africa nations spilled over into Somalia earlier this year when Eritrea sent at least three shipments of arms and Oromo fighters to Aidid. [. . .]
The move overnight Friday followed a drive-by killing Thursday of an Islamic fundamentalist commander in the capital Mogadishu.
Relatives said the family and officials of his Al-Itihad Al Islamia group suspected Ethiopian agents assassinated the commander, who in the past seven years organised raids to eastern Ethiopia's Ogadeen region, which Al-Itihad wants joined with Somalia.

On a more local level, at Moyale, the meeting of the security delegations at Nazareth one month earlier had some effect. About 1000 Degodia and 3000 Ajuran who had a recognized refugee status in Moiale, Ethiopia, were reported as going to be repatriated to the Wajir District of Kenya.[31] Both

---

[30] A few obvious mistakes concerning commas and quotation marks in this paragraph have been corrected.

[31] *East African Standard*, 8 July 1999.

groups had gone there as refugees from earlier clashes with each other. Nobody had thought about them for a long time, but as the 1999 census was approaching, office holders in Kenya who were keen on maintaining status of their administrative units needed their population to come back. In fact, movements of Degodia and Ajuran in both directions across the Kenyan/Ethiopian border have occurred for a long time and continue to occur.

As a consequence of the relaxed situation former sanctions against the Boran were lifted. Ethiopian Boran had not been allowed to sell livestock into Kenya. Now a livestock market was opened on the Ethiopian side in very easy reach of Kenyan buyers. This drained the Moyale (Kenya) market immediately of its supply, which in the past had largely consisted of cattle smuggled across from Ethiopia. As market fees are the principal income of the County Council, an agreement had to be reached. The Ethiopian officials proposed to divide that source of revenue by the nationality of the market actors. The mostly Ethiopian sellers would pay two Birr per head of cattle, and the Kenyans would be free to charge what they wanted on the buyers who were mostly from Kenya. For the Kenyans this was not enough. They wanted to spoil the Ethiopian market by imposing a prohibitively high transit fee for each cow taken into Kenya. They also increased the 'export' fee on each lorry taking cattle out of Moyale to the central Kenyan markets from 2,000 Kenya Shillings to 14,000 on the assumption that all the cattle were from Ethiopia. The *East African Standard* reported on 8 July 1999 that a crisis meeting was held at Moyale on this issue. Apparently the whole trade had ground to a halt.

The year 2000 brought continuous new variations on old themes. Violence escalated mainly around Moyale and Isiolo and the adjacent areas to the east of Mandera, Wajir and Garissa Districts, and double-digit numbers of casualties in one single confrontation became a regular feature of the news. Apart from the usual level of livestock rustling Marsabit District was relatively calm. Marsabit town had had its share of urban violence and clashes about agricultural land in the vicinity of the town in 1998, when there was some killing going on between Boran, Gabra, Rendille and Burji.

The main line of confrontation was that between Boran on the one hand and 'Somali' of various categories and degrees of 'Somaliness'. In Moyale some *Worr Dasse*, the 'people of the mats', i.e. the camel pastoralists or former camel herders who once had formed the *Worr Libin* alliance with the Boran as senior partners, now sided with the Boran again. The Ajuran whose gradual defection in the 1920s and 1930s we have described elsewhere (Schlee with Shongolo 2012) at some length, now resumed their position of the 1910s again, appealing to the memory of their old political affiliation on the side of the Boran. The Garre did not; they stayed aloof from such tendencies. They became the incorporation of the threat of Somali expansionism in the Moyale context. In fact, it was violent competition over grazing resources also with the Ajuran, which drove the latter back into the arms of the Boran. Confronted by other Somali, Garre and

Degodia, many of whom they had once helped to get access to resources then still under ultimate Boran control, the Ajuran fondly remembered the once privileged position they had at the side of the Boran.

In Isiolo the role of the 'alien Somali' was filled mainly by Degodia. There had been a fresh influx of them in the 1997 drought. There had been no place to go back to for these Degodia. The drought of 1997 was followed by the 'El Niño' floods of 1998 which decimated the weakened livestock. Since that deluge the drought had resumed. Vast areas of East and North East Africa had not seen a proper rainy season in 1999 and 2000.

In both settings, Moyale and Isiolo, in addition to pasture and water, conflict over resources associated with urban life and the national state played a role. From the fact that Meru, Bantu speakers from the neighbouring Meru District who form a business community in Isiolo, got involved in the clashes, it is clear that these were also about trade niches. In Moyale, in addition to recent pastoral and agro-pastoral Garre immigrants, there is also an old Garre business community which has been there from the very beginnings of the town.

Electoral politics are another field on which games of inclusion and exclusion are played. In 1997, immigrant Degodia were welcomed by the Boran politician Charfanno Mokku who then contested for the Isiolo North parliamentary seat, because they voted for him. Later they appear to have threatened the Boran majority in the electorate, and Mokku eloquently opposed their continued presence. In Moyale the same threat to the Boran majority of votes went out from the Garre. Mokku also appears to have played the ethnic and religious card in other ways. He referred to a rival Meru candidate, Kiome, as a 'non-Muslim Bantu'[32] showing some acquaintance with language families which are normally not part of popular identity discourses. In the words of the *Daily Nation*, 14 May 2000 (p. 8), the ethnic aspects of Isiolo electoral politics are the following:

> According to the Borana, the problem goes back to 1996/97 when groups of Somali herdsmen were allowed to move into the district. They came in search of water and pasture, as their traditional grazing lands in Wajir and Mandera were going through a nasty, dry spell. They were mainly from the large Degodia clan from Wajir, but there were also Murule clansmen from Mandera.
> One area from which they sought pasture is a place called Merti, a green enclave that is in the basin of the Uaso Nyiro River. Tensions with Borana would eventually develop over access to this basin. [. . .]
> There is the [. . .] opinion that, in fact, the 'Welcome' the Borana say they gave the Degodia in 1997 was politically calculated. As it were, the Meru presence in Isiolo has always been large and powerful. Indeed, a couple or so Meru MPs have been elected in the district, including the man Mr Mokku defeated in 1988, Mr Muthaura Kiome.
> As is the case with their kinsmen, the Merus of Isiolo tend to vote Opposition ever since pluralism set in, their party being the Democratic Party. Politically, it made sense in 1997 to allow an influx of Somali who, like the

---

[32] *Daily Nation*, 14 May 2000 (p. 8). Francis Muthaura Kiome, incidentally, later represented the Degodia as an advocate when the Wagalla massacre of 1984 finally came to court in Nairobi in 2007.

Borana, are Muslims and who instinctively vote KANU. That way, the Opposition challenge would be diluted and even neutered.

Mr Mokku was the beneficiary of the immigrant Somali KANU vote in 1997, though not without a fierce fight from one Tache Gaji, a fellow Borana who challenged him on a DP ticket and was fully backed by the immigrant Merus. At the time, Mokku could bask in the glow of having helped, or at least acquiesced, as the Somali came over. The Somali clans gratefully paid back the favour with their vote. The marriage was not fated to remain blissful for long.

The close interplay of local politicians, elders (in the sense of representatives of local agro-pastoral communities or 'elders' hired by politicians as democratic fig leaves and amplifiers of their positions), administrators (often more as instruments than as actors), and the media is very well illustrated by the Moyale case.

In the first phase the violence took place mainly on the open range, in the grazing areas in Wajir District and in the eastern parts of Moyale District. In March 2000 there was a disagreement between Garre and Ajuran about the appointment of a chief for El Danaba location and a sub-chief for Irrees T'eenno. The area had hitherto been dominated by Ajuran. Now the Garre claimed these offices.

A Garre was killed near Gurar on 16 March 2000. Then a mine exploded at D'okisu in Ethiopia and killed 12 Garre. The Ajuran were suspected to have planted it. On 22 March one Ajuran was killed by a Garre near Gurar. Then two Ajuran hamlets were burned. On 27 March four Garre were killed at Qofoole in Wajir District. Then the killings spilled over into Moyale District. A lorry, a Canter, belonging to an Ajuran businessman was attacked on the Moyale–Dabel road. There was an exchange of fire between the escort of the lorry and the bandits who were believed to be Garre. The same evening two Garre were killed at Nana and their genitals were removed. One of the bodies, that of an elderly man, had an Ajuran property mark carved into its thigh. Garre now started to flee into Ethiopia while Ajuran sought safety to the south and moved in the direction of Isiolo. Killings continued on this scale throughout early April. In mid-April there were major clashes in Moyale town itself.

On 4 April 2000, there was a demonstration of Garre in Moyale town following the arrest of two elders who had been accused of fuelling clashes between Garre and Ajuran in the north of Wajir District. Demonstrations and riots continued the following days. In the course of these demonstrations one Kenyan flag was burned on 5 April 2000, an incident which gave rise to rumours that the flag of Region 5, the Somali region of Ethiopia, was raised in its stead. Under the pressure of these demonstrations a cash bail of 30,000 Kenya Shillings each was accepted for the two elders on the initiative of a Mandera politician from the opposition Ford Kenya party, a Garre, who had come with a lawyer and a group of human rights activists. A young Garre was shot by a policeman on 7 April 2000, as a riot developed between the Garre demonstrators, and Gabra and Boran youths. The Mandera politician was informed of that and came again, this time with a pathologist to perform a post-mortem, newsmen and human rights activists in an aeroplane. The group was prevented by Boran and Gabra, in collusion with the District Commissioner, from proceeding from the airstrip to town. The next day stone throwing between mobs from the two sides continued and shops were looted. On 10 April 2000, Dr Guracha Galgalo, the Member of Parliament for Moyale and an Assistant Minister for Health, arrived on the scene. At a meeting with Gabra and Boran elders the following day he explained:

Nu lafa teenna ta waraani itti bu hin dandeennu. Lafa tana nami duri Waaqi itti d'ale Gabraafi Borana. Sakuyeen Boranuma. Gaafa isan na filatani, ka anin olla chuf keesa deeme jaaroleen haasau, jaarole Borana, ta Gabra, ta Garri, ta Sakuye, ta Burji chufa haasaeeni.

We cannot tolerate that the spear comes down on our land. Long ago, the people whom God begot on this land are the Gabra and Boran. The Sakuye are just Boran. When you elected me, I went to all villages and talked to the elders, the elders of the Boran, of the Gabra, of the Garre, of the Sakuye, and of the Burji, all of them.

Guyya tokko mana nama beesharaa keesatti jaarole Garriitin haasae. Guyya suni jaaroleen Garri tokko alaati hin hamne. An akkasi jed'eeni:
Waan siasa barbaad kura keesatti wal d'amne, namu ka bira deemu.
Amma maendeleo *ijaarsa lafa ka waliin duunnu. Isan gaafa kuraa nama d'aabatani qabdani. Amma ammo* kura ya naa bue. Isaniillen *namu ka kura naa d'aeefi ka naa hin d'aini kamu amma kiyya. [Nami] keesani gaafa kura fed'atu dubbi jabdu jed'e. Waan jed'e 'D'adac Warraabi akaaku keenatti fuude, D'adac Hariyyaa, abbooti teenatti fuud'e, ka Obbu amma nutti fuud'a' jed'e.*

One day I talked to Garre elders in the house [of a Garre businessman]. That day none of them remained outside. I said to them like this:
We disagreed on matters of politics during election time, let everybody leave that aside. Now let us come together for the development [in Swahili] and the building up of the country. During the election you had your own candidate. But now the election [in Swahili] has turned out in favour of me. As of you, the ones who voted for me or not are now all mine. Your [name of candidate] said something strong during the campaign. He said: 'Our grandfathers took the Acacia of Warraab [in Somalia], our fathers took the Acacia of Hariyya [in Ethiopia, near Lae, a short distance North East of Moyale], and it is us who will take the Acacia of Obbu' [near Sololo, in Kenya, beyond Moyale to the west].

Dubbi suni jecum siaasa jed'ani kara rahiisatti fud'atani. Ammo nama wa hubatuu jeci suni yaada fagoo nama garsiifti. Gaafum sulle Ali Bod'een akkasi jed'u suni, D'adac Obbu ka fud'atani d'iisi gaara kan irra kura fed'ata gadi himbuune.

That statement was said to be just politics and was taken lightly. But for perceptive people it showed a long perspective. Even at that time, the very [name of the man] who said so, let alone taking the Acacia of Obbu, never even came down the hills to campaign for the votes here. [I continued to speak to them like this:]

Ani Sololotti d'aladd'e. Finni armaatifi ka Sololo diqqo wal hanqata.
Nu warri Sololo guddo naga feena. Nageenni ammo kiwango *qaba.*
Yo atin naga feete ka nami isan waliin jirtani naga hin feene, guddo shida.

Lafti tunin ta Gabra, ta Borana ta Saakuye. Isaniillen, duri taanan, amma d'io taanan, ya d'uftani qubattani. Ya warritani. Marra bisaan waan lafa tan jirani walumaani sooranne. Obboleeyan ta waliin d'alateellen wal hangafti. Angafi lafa tana Gabraafi Borana. Amma wanni anin isan kad'add'u nagaan taa, isan kamu raaya Kenyatti. Lafa tana ka walumaan jaaru.

I have been born at Sololo. The customs of here and those of Sololo are a bit different. We people from Sololo love peace very much. But peace has a limit [in Swahili]. If you want peace and the person with whom you are together, does not want peace, it is a big problem [in Swahili]. This land belongs to the Gabra, the Boran, the Sakuye. Also you, a long or a short time ago, have come and settled here. You have become part of the family. Pasture and water which are in this land have been consumed by us together. Even brothers are senior to each other. The first-borns of this land are the Gabra and Boran. But I ask you to stay in peace, you, too, are citizens of Kenya. Let us build this land together.

[. . .] Nami arra ammo lafa namaan falamu, nama naga lafa tana balleessa barbaadu. Gaafi falamani arraniti. Nami lafa tana ta ufiiti herregatu, naga lafa tana balleessuu himmalle.

Somebody who wants to claim this land today is the one who wants to spoil the peace of the country. This is not the time to make claims. Whoever depends on his use of this land should not spoil its peace.

Waan anin seu Garriin lafa tanaati d'alatte ta oobru qabdu, ta mana d'akaa jaaratte qabdu, yo lafa tana falamaa arra kaate isiittu naga hin feene. Ammo akka anin seuuti nami arra falama tana ol ejju, keesumma, namum kale lafa tana seene, ka ojja nageenni lafa tana hammaate ufi wa saamu barbaadu.

I think that the Garre who have been born in this country and own farmland and own houses built of stone, if they rise to claim this land, it is themselves who do not want peace. But I think that somebody who is a guest and now rises to make such a claim, somebody who has just come yesterday to this land, just wants to rob something when the peace has been spoiled.

[. . .] Warra naga kana hin feenne, wom diqooti irra bada. Yo lafti tuni naga hin qabne fula jiru sulleeti woma dandae hin hojadd'u. Jaaroleen akka Haji Raago fa, Haji Mamad Hapicha fa, kamu qasaara guddooti egeri qabata yo lafti tuni egeri naga d'abde. Besharaan warraallen egeri himbaddi. Ta namuutu akkasum baddi.

The people who do not want peace, lose very little. [But] if there is no peace in this land, I cannot do any work at my place [i.e. Nairobi]. Such elders as Haji Raago and Haji Mamad Hapicha [rich Gabra businessmen], each of them will suffer heavy losses if peace gets lost in this land. These people will lose their business. Even others will lose in the same way.

Jarri obboleeyan teenna ta Garrii lafa ufiiti naga qabdi. Beesharaan warra deemutti jirti lafa Maderatti, ta Wajeeratti. Teenna ammo arra ya baba-doofte. Amma wanni nu d'umneefi akkamiin naga lafa teenna deefanna, tanaafi d'umne.

Those brothers of ours of the Garre have peace in their land. Their business will go on in Mandera and Wajir Districts. But ours has even been disturbed today. But now we have come to find ways how to return peace to our land; that is what we have come for.

69

Lammeeso, yo warri Moyale wald'abe, nami lafa d'ibii ka akka [nama siasa Mandera], d'ufe naga lafa tana booressuu him malle. Ammalle dubi bandera Kenya fa gubani, yo nama bandera gube, OCPD betrol itti dare guba jed'e, sun teennaniti. Yo sirkaali bandera isa ka gubani dubbi irra hin qabne, tun shoori isa.

Second: if the people of Moyale quarrel among themselves, it was not right for somebody from another land like [the politician from Mandera] to sully our peace. And as to the affair about the burning of the Kenyan flag, if somebody burns the flag and the OCPD [Officer Commanding Police Division] poured petrol over it and said 'burn it!', that is not our affair.[33] If the government does not worry about its flag being burned, it is up to them.

Nu ammo ta nu d'ibde biyye teenna irraati bandera gubani, haganumaati nu d'ibe. Nu bandera sirkaal Somali himbeennu, ka Topia himbeennu. Ka nu naga jalaati arganne eegi Kenya uhuru argatte, ka amma biyye teen iraatti guban kana. Ammo akka nu seenutti abba dubbatuufi hin qaba, tan asum lakimne duubatti chaqamna.[. . .]

But we are disturbed by the fact that the flag was burned on our soil; that is what worries us. We do not know the Somali flag, nor the Ethiopian one. The one under which we have found peace ever since Kenya got independence, that is the one which was burned on our soil. But we think somebody will take up this matter, let us leave it for the moment and see what happens in the end. [. . .]

On the following day, 12 April 2000, an aeroplane was expected from Nairobi bringing journalists and the lawyer hired by the Garre. A delegation of local politicians and businessmen wanted to stop these people at the airstrip to avoid negative press coverage for Moyale. Among them was the MP, Dr Guracha. The plane also brought the newly appointed DC, Clement Nzoimo Kiteme, and Ali Barricha, the Ford Kenya official from Mandera. The incoming DC was introduced by Barricha to the area MP and at the other leaders on the airstrip. That is how they first heard about the fact that now they had a new District Commissioner. They were infuriated by the fact that the new DC was introduced to them by an opposition politician and concluded that he must be pro-Garre and anti-KANU. The controversy even led to blows between the County Council chairman, Mr Golicha Galgalo and Mr Barricha.

Mr Barricha was alleged to have handed over a newspaper to someone in which a map labelled 'Garriland' [sic] was found which shortly led to bitter controversies for, as we shall see, totally absurd reasons.

Following the news of the DC's transfer, the town talk was that the

---

[33] Of course, the police commander was not present at the burning of the flag and did not assist in doing so. But Moyale grapevine had it that there was a disagreement between him and the District Commissioner about the arrest of the two Garre elders. He predicted problems and rightly so, since about 2000 Garre converged on the DC's office the next day, throwing stones at it. The OCPD, these rumours claim, would have preferred to accept a bribe provided by the Garre business community which he wanted to share with the DC, but the DC had insisted on keeping the elders under arrest. When the DC's office was stoned, the OCPD is said just to have looked on, forcing the DC to call in nearby army units with tanks. This latter action found a very negative response in the press and may have contributed to the DC's subsequent transfer and later forced leave.

Garre community had overpowered the so-called indigenous community by being able to bring in a new DC of their choice, alleged to be a Ford Kenya DC. The outgoing DC, Mr Kipkebut, had been accused by the Garre community of siding with the Gabra/Boran.

During the week from 7 April to 13 April 2000, the Kenyan/Ethiopian border was closed at Moyale by the district administrations on both sides, after two days of joint military operations, to prevent Ethiopian Garre from assisting their Kenyan tribesmen who lived in an area directly adjacent to the border. Because of the economic interdependence of this border area, the economic effects of this measure were immediately felt on the Kenyan side. The price of a tin of maize went up from 13 to 20 Kenya Shillings (Ksh), that of a kilogramme of sugar from 40 to 60 Ksh, a cup of milk from 8 to 20 Ksh, a bundle of miraa (*qat, chat*) from 20 to 35 Ksh.

A curfew was also proclaimed in the Kenyan town of Moyale from 1800–0600 hrs, a directive which reminded the older generation of the 1960s and 1970s when Moyale was under curfew for about eight years. The children born in that period are still known as *ijolle kafio*, the 'children of the curfew', and said to be particularly numerous, because husbands had nowhere to go at night.

A Garre perspective on the new frontlines in these ethnic clashes is provided by the following interview Abdullahi Shongolo had with the councillor Adan Sheikh, the County Council member for Nana location,[34] on 14 April 2000:

> Q. *Waan ani si gaafadd'u Garriifi Borana maanti arra wal d'absiise?*
> What I ask you: what has caused the quarrel between Garre and Boran?
>
> A. *Waan duri jiddu teen baeeti arraalle bae.*
> What has come up between us before also has come up now.
>
> Q. *Waani kuni maani?*
> What is that?
>
> A. *Saapuppa tokkooti jira, ka jiddu teen jirtu, kanaati waltiin nu yakke. Beeta niitin garooba ta d'iirsa yakkite bira deemte, ka itti d'ufteelle hin mid'aasitu. Borani qara woma hin hubatu, niitin garooba tanaatu balleesa hin agarra.*
> There is one spider web between us which has brought us up against each other. You know, a divorced woman who did wrong to her husband and left will not do any good to the next one. The Boran normally do not understand things [quickly], we shall come to see that this divorced woman will cause damage to them.

Both the 'spider web' and the 'divorced woman' are a reference to the Gabra [Miigo]. The 'spider web' is meant to describe their weakness and futility, while the 'divorced woman' refers to the circumstances in which previously the Gabra sided with the Garre, from whom they might have

---

[34] A bit over five months after the interview, the night of 24 September 2000, Adan Sheikh was murdered by unknown gunmen outside his house as he came home from the mosque. (*East African Standard*, 26 September 2000 p. 4)

separated over some quarrel about the division of the loot, but now they side with the Boran.

> Q. *Warri kuni daba isan dabsitaniifi Boranaatti d'eete, amma lakiftani ho re?*
> These people withdrew to the Borana because of the bad things you did to them, so why don't you leave them alone now?

> A. *Hin lakimna ammo warriillen ka Islaanan walti nu naqu lakkisu.*
> We let them alone, but they should also stop bringing dissent between us and other Muslims.

> Q. *Warri Boran mo Islaanan walti isan naqe?*
> Did they make you quarrel with the Boran or with other Muslims?

> A. *Boraniilen amma Islaanuma ammo Ajuranaan walti nu naqe. Beeta, Gabarti amal iyyeessa qabdi. Ati ilmaan iyyeessa ta guddatte take garte? Ilmaan iyyeessa yo waan isiin feetu chufaalle itti kennite, hammeennuma sii qabdi. Miisholle isiin doota. Warri kuni akkasi yoosi. Yo sila isiin nuufi Borana jiddu hin jiraanne, nu Borana wal hin d'abnu.*
> Even the Boran now are Muslims, but they made us quarrel also with the Ajuran. You know, the Gabra [using the Amharic near homonym and purported synonym *gabar* which stands for dependents of low status] have the habits of poor people. Have you ever seen the children of poor people who have grown up? If you give everything they want to the children of the poor, they will still treat you badly. In the end you will die because of them. Now, these people are like that. If they were not between us and the Boran, we would not quarrel with the Boran.

In the evening of 12 April 2000, there was a report on BBC radio. It is claimed that in this report it was said that 90 per cent of the population of Moyale District were Garre. A local Garre, a private telephone operator, was suspected to have given the BBC an interview which contained this information. He was called to the Security office and admitted to that.

In the resulting atmosphere of increased sensitivity, the matter of the map which had been found between the pages of a newspaper which Barricha had given to someone at the airstrip came up. Photocopies of it were sold in several shops for 50 Ksh a piece. It was a completely innocuous map presumably from a book by Shun Sato (Sato 1996 p. 276), which is meant for rough orientation to show whereabouts the Garre live. Many other ethnic groups are mentioned as well in their approximate locations. The legend clearly states that the uninterrupted bold lines on this map are the main roads. Nevertheless there were heated discussions in which the roads leading from Dila in Ethiopia to Dolo (where the borders of Kenya, Somalia and Ethiopia meet), to Wajir and up again via Moyale and Mega to Dila were pointed out as the boundaries of 'Garriland'. It was said that the Garre claimed the whole area comprised by this triangle as theirs. The literate ones among those who forwarded that interpretation must have known that they were lying. [Georg Haneke, a student of Schlee who was in Moyale in June 2000, still found people debating this map. He pointed this misreading out to them.]

This alleged territorial claim by the Garre led to heated comments. In a meeting attended by selected elders of Garre, Gabra, Boran, (the District Security Committee = DSC team), and delegates from Nairobi led by the MP Dr Guracha, a Boran councillor gave the following speech:

Nami odu Moyale tana ala chaqasu,waan Gabra Boranaafi Garriin wal d'abdu sehani. Ammo wanni wald'abaaniti, falama lafa. Shahidi falama tana waan hedduuti garsiisa. Banderan chirani falama lafa garsiifti. Lafa dira Moyale ka gargari kutani, gama tokkooni kilili shani jed'ani, falama takka hin argini garsiifti. Lafti il boruu ta Gabra Borani keesa him baane, falama garsiifti. Tun chufa waan nu garre. Amma ammo takka hin garreen ya nutti d'ufte. Tun tami? Mepi mid'aasani lafaan falamani. Ta tana [jiifatti] isanii diiga.

For those who listen to the news of Moyale from outside, it appears as though the Gabra, the Boran, and the Garre are quarrelling. But what they are quarrelling about is the claim to the land. The evidence for this claim to the land is shown by several things. The flag which was torn down shows the claim. The urban land of Moyale which was divided, one part of it being said to belong to Region 5 [of Ethiopia], demonstrates this unheard of [literally: unseen] claim.[35] The land to the east where Gabra and Boran cannot step in shows this claim. We have seen all that. But now something we have never seen has come to us. What is that? A map was made to claim the land. As to this, [mentions the name of a former chief] will explain to you.

Former chief:

Akka Kaansolaan [. . .] baanu, olkiin arra Moyale keesatti baate, wald'aba gosaatiniti, falama lafa. Garriin gaafa demonstration *tariqi shani, waan jette, 'lafti teenna male teesaniiniti' nuun jette. Ato tanum itti harka wal qabnu, Waaqi d'ae nuttiin d'ufe waan shahida falama lafa tana kuwanja endegeeti, mepi nam irraati argani. Mepi suni waan garsiisu lafti tuni 'Garri-land' jetti, lafa Garii jecu.*

As Councillor [. . .] has said, the war which has now come up in Moyale is not a tribal quarrel but one about claims to the land. During the demonstration on the fifth the Garre said to us: 'this land belongs to us and not to you.' As we were at odds over this, God brought us the evidence for the [fact that the Garre] claim this land to the airstrip where the map was found with somebody. What this map shows is said to be 'Garriland', the land of the Garre.

Mepi suni yo nu laallu, Kenya keesa qara maqaan Boranaatu hin jiru. Yaballoofi Mega jiddu maqa Borana arti kaani. Seera Somali, Topia keesa, Waachille fuud'ani Moyaleetti d'aabani haga laga Waraabessa. Arti Bunatti d'aabani. Artiin Doolotti d'aabani. Dooloo- Waachilleeti d'aabani. Mepi suni kana.

If we look at this map, inside Kenya the name Boran does not even appear. The name Boran is put between Yaballo [Yabelo] and Mega. The Somali [Region 5?] boundary inside Ethiopia goes from Waachille via Moyale to Lag Warabeesa [near Turbi in Kenya, pointing to other roads]. From here it is drawn on to Buna, from there to Dolo. From Dolo it is drawn to Waachille. That is the map.

---

[35] The border town of Moyale on the Ethiopian side is divided, the western part belonging to Oromia, the eastern part to what was then Region 5 of the Ethiopian Somali, since renamed Somali Regional State.

The map from the book by Shun Sato (1996) has been reproduced in Schlee and Watson (2009 Vol. I p. 212). A look at either of these books will convince the reader that there are no internal boundaries of any state at all on the map. There is not even any mention of the Somali area of Ethiopia. The ex-chief continues:

> *Yo Borana Garriin asi wald'abde, Jaarsi Gabra Garrii, ka Burjii Borana lafa tana hin jira.* Mujumbe *hin qabna. Wanni warri siasa, ka Madera asi gadi maruufi, ka akka* bunge *Maderatiifi [warri d'ibiin] as maruufi,* madai *lafa tana irre itti tocaniifi armaan marani.*

If the Boran and Garre now are quarrelling, there are elders of the Gabra and the Garre, of the Burji and the Boran in this land. We also have a member of parliament. The reason why the politicians from Mandera come here all the time, like the MP for Mandera and [others], is that they want to strengthen the claim [in Swahili] to this land.

> Mepi *kana [warra siaasatti] d'ufeeni, isaati deemani. Falama lafaatifi deemani. Garriin Moyale* mepi *kanaan lafa hin falamne. Ammo yo wal qub qabu d'ibaataniille,* mepi *kunin haga lafa onana killili shani jed'aniini ya itti* shahida *bae. Puruufi ya nuu mullate.*

It was [a politician] who brought this map, he brought it along. He brought it for the claim to the land. The Garre of Moyale have not claimed the land with this map. But whether they are aware of each other or not, this map supports the fact claim that they put forward the other day that this land belongs to Region Five. The proof [in English] has appeared to us now.

Readers with some knowledge of Boran and Swahili will have noticed that there is a degree of code-switching in these speeches. In some places, where these changes of register appeared rhetorically significant to us, we have highlighted non-Boran words by rendering them in a different type. This may help finding them, because in the text they appear in various degrees of phonetic assimilation to Boran and may not be immediately recognizable.

In the preceding paragraphs *madai* (claim) and *shahida* (testimony) are from Swahili. The use of Swahili words here may be a signal that the speaker is familiar with the sphere of statehood and government and that the matters he addresses are of legal importance beyond the local context. The speech therefore should not be taken lightly. That the English word 'proof' has been used seems to serve the same purpose.[36]

Whatever the rhetoric strategies were, they were effective. Pressure on the Garre delegates during this meeting became so strong that they had to apologize for everything that was alleged against them, including the plainly absurd allegations involving this map, and had to pay 5,500 Ksh for appeasement (*hoola*) to the Boran and Gabra. The meek response by a

---

[36] Other words like *mjumbe* (Sw. for Member of Parliament), *bunge* (Sw. for parliament) or *mepi* (English: map) apparently have been used simply because in Boran there are no words of Oromo derivation for these institutions and things. In Kenya, Boran Oromo has no official status and there is no language board inventing neologisms.

Garre delegate only appeals to some common, community oriented activities in the past.

A Garre elder (one of the two liberated a couple of days before):

Nu naga barbaanna arma d'umne.Waan guddo wal himanu hin qabnu yo naga barbaan jenne. Obboleeyan teenna ta Gabra maqa kufrimadi nu himatte. Madarasa Alhuda waliin jaare. Masjid waliin jaare. Imaamkin Maalin Isaaq, Gabra. Yo akkasi akkamiin kufaar jennaani, nuun chufa walumaan Isilaana?[. . .].

We have come here to look for peace. We do not have much of which to accuse each other, if we have come to look for peace. Our Gabra brothers have accused us of calling them infidels (*kufaar*). We have built the Qur'aan school Alhuda[37] together. We have built a mosque together. The Imaam is Maalim Isaaq, a Gabra.[38] It is like this, how can we call them *kufaar*, as we are all Muslims? [. . .]

Gaafa barana Funaan Nyaatatti nama fit'ani, nama gosa hedduutti keesa jira. Gaafasu obboleewan keen ka Gabra nuun hamatte. Yaamne gaafanne. Yo Dogodiaan wa yakkite, maanifi walti nu hamatan jenneeni. Gaafasu nuufi Dogodiaan walti nyaapa. Duraanu hori keenna, gaala kuma lamaatifi d'ibba saddeeti nurra fud'atte. Nama haga afurtama nurra ijjeefte. Dogoodin nuu nyaapa. Gaafasu ammo Dogodifi Borani obboleessa. Nulle akkuma biyya haatu nu haati [. . .].

At the recent massacre at Funaan Nyaata there were many tribes involved. At that time our Gabra brothers talked badly about us. We called them and asked them. If the Degodia have caused harm, why do you blame us together? we said. At that time we and the Degodia were enemies of each other. They had taken our livestock, 2,800 camels, from us. They had killed about forty of our people. The Degodia are our enemies. But at that time the Degodia and Boran were brothers. As to us, they attack us the way they used to attack other people [. . .].

An onana gaaladiin na qabde. Ani jaarsa Garrii challaaniti, ka Gabra Borana, ka Burji, jaarsa laf tana. Isani chufa boole irra d'abe. Namaati narra due. Awwaal isan irra d'abe. Jifu duri wal irra qabnu chufa isan irra d'abe. Waraan natti kaasitani. [. . .] Waan ijoollen duuriyyee faan taatu, gosaan himtani. Dubbiin killili shani arma hin jirtu. Warra gaaseta, nu hin yaamne.

The other day I was arrested by the pagans [meaning the Kenyan Government]. I am not only an elder for the Garre, but also for the Gabra, Boran, for the Burji, an elder of this land. None of you expressed his sympathy with me. Somebody of my own has died. [A reference to the young Garre shot by a policeman on the 7th.] None of you came to express condolences [Literally: I missed you at the burial]. You missed to carry out all the obligations which we had towards each other. You raise the spear against me. [. . .] You cannot blame on the whole tribe what some hooligans did. There is no talk about Region 5 here. And the newspaper people, we did not call them.

[. . .] Gosa teenna takk hin garre nutti baasitani. Nama yo inni due, lazim hin awwaalani. Maanin dueelle barbaadun haqqi. Amma ammo lafti teenna

---

[37] According to some Gabra, Al-hu-da stands for three donors who are all Garre, namely Ali Bod'e, Hussein Dakha, and Dauud Intallo. This does not point to Gabra participation.

[38] It is said, however, that there are Garre who refused to pray behind this Gabra Imaam during this conflict.

irraati hin awwaalan jettani. Yo endegeeni reefa biimiti d'uftu, dura ejjatani d'owartani. Takka sidiillen akkana wal hin tocine. Awwaal nu waakattani. Arra nu taasiaati harka nu qaba, isan ammo dubii waraanan harka nu qabdani.

[...] Our tribesmen, you have brought forth what we have never seen. If a man has died, it is necessary to bury him. It is also necessary to find out of what he died. But now you say that the land belongs to you and he cannot be buried there. When the aeroplane came to examine the body, you blocked it. Not even enemies have ever done such a thing to each other. You deny us the burial. Today we are in the mourning period, but you are bothering us with a case and with the spear.

Borana Garriin soddaa obboleesa. Boran jifu lencha waraabessaa kennu, nu arra ka soddaa obboleesa jifu aada nama irra d'abne. Bineesa alaalle hin hanqanne. Obboleewani, me nama narra awwaala, duubatti dubbi qabdani baasa.

The Boran and the Garre are affines and brothers. From the Boran who give their due even to the lions and the hyaenas, we have today missed our due as affines and brothers. We have been degraded below wild animals. Brothers, help us to bury our dead and bring forth any case you have afterwards!

Another Garre elder added:

Nu nagumaaf arma d'ufne ammo goci teen ta Gabra Borana statement *kotini nuu soomte, torbaan tokkoot laf tan irra ka'a jette. Waan nu d'ageenu, nami* notice *torbaan iti keenanif,* ambassador *challa. Nu* ambassandaanit, *worr qubate laf taau. Nu oboleyaan teen tan irra hin éeganne.*

We have come here for peace, but our tribesmen of the Gabra and Boran read a court *statement* to us, and told us to leave this land within one week. What we have heard, is that a seven days' *notice* is only given to an *ambassador*. We are not *ambassadors*, we are people who have settled in this land. We have not expected such a thing from our brothers. (The terms in italics are English in the original.)

The paper which is here ironically referred to as a court statement and which is compared to the notice to leave given to a diplomat who has become *persona non grata*, is the 'Declaration of Indigenous Communities' analysed in the next chapter.

On this occasion, the Garre elders were under time pressure to get their dead buried. Being aware of this situation, the Gabra and the Boran thought that it was an opportunity to corner them and to pressurise them to accept guilt for having caused violence in Moyale District. The Garre were unaware of these tactics. Thus, in the end, the Garre elders accepted guilt and apologized to all by offering *hoola*[39] as a token of goodwill (the above mentioned 5,500 Kenya Shillings) in order to conclude the matter.

The pressure on all actors on the Moyale scene and the heat of the controversy were increased by the circumstance that simultaneous

---

[39] B. *hoola* is equivalent to Som. *subeen*. It literally means a young female sheep. It also stands for the gift by which one asks for forgiveness in an attempt to open negotiations about compensation to avert revenge.

to these events there were clashes between Garre and Ajuran in Wajir District and between Garre and Degodia at Malka Wiila near Dolo in Ethiopia where 37 Garre were reported dead. On 12 April 2000 there was a demonstration in Wajir by Degodia, Ajuran and Ogadeen Somali who demanded that the Garre should leave that district. There were killings and the Garre fled towards Moyale. On 17 April 2000, seven lorries full of Garre refugees arrived at Moyale. That did not ease the situation there.

In the *East African Standard* of 26 June 2000 (p. 8) Abdikadir Sugow reports:

'DC denies evicting Garre clan members'

Embattled Wajir District Commissioner, Mr. Fred Mutsami, yesterday denied having forcefully evicted hundreds of Garre families from their homes in Bute subdistrict.
[. . .] It was members of the community who requested to be relocated to the nearby El-Danaba trading center for fear of attack from rival Ajuran clan. The DC further explained that the Government had provided vehicles and security for the community during the relocation exercise. [. . .]
About 5,000 members of the Garre clan, among them 720 school children who are currently starving at El-Danaba, have accused the DC for instigating their eviction while supporting their rivals. [. . .]
El-Danaba being a volatile area, where several Garre were already killed, was not a safe place for the Garre to be relocated. Instead they opted for Moyale where they arrived by night and settled among their own kinsmen.

The eviction exercise was said to be a strategy employed by the government to secure lasting peace to the areas affected. Earlier, in Isiolo, two Somali clans were to be repatriated from Isiolo District. The *Daily Nation* newspaper of 26 May 2000 (p. 48) reports:

'Clans ordered to leave Isiolo'

Two Somali clans must leave Isiolo for lasting peace to be achieved, Cabinet Minister Marsden Madoka announced after meeting the town's leaders. The Minister said the government would repatriate Degodia and Murule clans people in line with the strong recommendation of elders from Isiolo communities [. . .] in a nine-hour security committee meeting. During the meeting, Mr. Madoka separately summoned Herti, Issak, Degodia, Murule and Ajuran Somali clan elders and also those of the Boran, Turkana, Meru and Samburu tribes. Mr. Madoka said that except for a few elders, all others agreed that peace would not be restored in the district until some sections of the Somali community were moved out of the area. [. . .]
Reacting to this order, the local Somali leaders had this to say after the meeting:
[. . .] Councilor Omar Elmi, Mr. Mohamed Ibrahim, the chairman of the Somali development Committee Mr. Farah Awad, and Isiolo Somali elders committee chairman Ibrahim Bile, members of the two clans said they would not leave Isiolo District under any circumstances. Mr. Elmi said: 'They would have to kill all of us and carry our bodies away'. Mr. Awad and Mr. Elmi said the Degodia and the Murule had settled and were grazing on Herti and Issak land and the Borana have no right to demand their eviction. They accused the Borana of 'intimidating everyone with their numbers'.

[. . .]
Mr. Madoka said he would go back to Isiolo next week 'or somebody much more senior than myself will come to announce of the modalities of the removal of these clans'.

The claim of the Somali elders Awad and Elmi that the Degodia and Murule had used land belonging to other Somali clans, namely Isaaq and Herti, and therefore no Boran interests were affected, goes back to colonial land allocations. Isaaq and Herti (mostly Dulbahante) from what then was British Somaliland in the north served in large numbers in the King's African Rifles and the Administration Police. Veterans of these forces were given grazing rights in the Isiolo District.

What this attempt to evict the Degodia from Isiolo District shows is how deep the division between the Boran and the Degodia had become again since the years of 'brotherhood' in the 1980s. The Ajuran mirrored this development. As the Degodia quit (or were expelled) from the Boran fold, the Ajuran tried to enter it again. To witness this process we have to turn to the Moyale scene again.

BORAN-AJURAN ALLIANCE

In a meeting on 9 July 2000, attended by representatives of Ajuran, Boran, Gabra, Burji, Sakuye elders and two MPs, Dr Abdullahi Ali (Wajir North) and Dr Guracha (Moyale), the chairman of the Moyale County council, Golicha Galgalo explained:

> [. . .] *Arra finnum qara nuun Gabra Borana, Sakuye Burjiin harka waliin qabnu, obboleeyan teen ta Ajuranaallen nutti daramte. Waraanum harka nu qabuuti warraalle harka qaba. Nyaapi waliin harka nu qabu tokkum, tanaafi finna walti d'umne. Finni nu walti mid'aafannu ka waraanatiiniti, ka nagaá, ka marra bisaani. Akkanaan me ka waliin falannu.*

> Now we, the Gabra and Boran, Sakuye and Burji, hold the old customs in our hands together, and our brothers of the Ajuran have also joined us. The war ['spear'] which has us in its grip is also affecting them. The enemy who puts pressure on us is the same, that is why we have come together to organize our way of life. The things we have to organize are not about war, they are about peace, about pasture and water. Let us plan matters along these lines.

Of course, there is an element of rhetoric in this. Pasture and water being the disputed resources, talk about them is not just talking about peace rather than war. Golich Diima, a Boran elder, added:

> *Eegi gada Sora Hoffoole, bunge qara, nu takkuma kora finnaati Ajuranaan walti hin d'umne. Abbootin teen waan duri nutti himte, goci laf tan quttuma walti aantu sadi jette: Gabra, Borana, Sakuyeen takka. Tun abbooti lafa. Goci nutti aantu ta finniillen walti nu duudu duri Ajuran jette. Abbootin teen Baalad jettiin.*

> Ever since the *gada* period [here meaning the term in Parliament] of Sora Hoffoole, the first Member of Parliament [for Moyale after Independence] we have never come together with the Ajuran to discuss the management

of our affairs. Our fathers have told us before, that the tribes who live in this land are three: the Gabra, Boran, and Sakuye together. They are the owners of the land. The tribe next to us whose customs were interwoven with ours are the Ajuran, they said. Our fathers called them Baalad.

Gaafasi Borani Waaso tan hin qubanne. Galaan Waaso kan maqaan gaafas Takkaari Qalla. Gaafasi Borani ka adamooti aci d'aqa. Adamooti, laf diida ta bisaan yaaán garte. Eegi chin Tiite Guraacha, Wajeera itti galte. Hariyaan gaafasi Wakor Mallu [. . .] Aabbo, Mamad Ali Gababaat d'ale abbeeruma. Gashe Ali Diima tana [. . .] Nu Ajuranaan akkan olla obbolees. Gaafasi, laf Wajeera tan Borana Baalad challaati qubata. [. . .]

At that time the Boran did not settle along the Waso [the Ewaso Nyiro River, in what is now Isiolo District]. The Waso River at that time was known as Takkaari Qalla. At that time it was Boran hunters who used to go there. The hunters saw an even land with water flowing through. After the disaster of the Black Flies [in the 1890s, these hunters] moved to and from Wajir. The *hariya* age-set[40] at that time was Wakor Mallu. [. . .] My father was begotten by Mamad Ali Gabaaba as a genitor [i.e. he was not begotten by his father but by an Ajuran lover of his mother]. He was of these Gashe [an Ajuran section[41]] of Ali Diima. [. . .] Thus we and the Ajuran were like one village and brothers. At that time at Wajir only Boran and Baalad [Ajuran] were settling. [. . .]

The retired Chief Huqa Guleid, a Boran, then elaborated:

Nu duri, sirkaalat eegi uhuru argate, Ajuranaan gargari nu baase.[. . .]

Long ago, after independence, it was the government who separated us from the Ajuran. [. . .]

From what we have described in another volume (Schlee with Shongolo 2012 Chapter 1 *'Pax Borana'*), it is evident that this was not exclusively the doing of the government and that the government in question was the colonial government, over three decades before Independence. Chief Huqa continues:

Ajurana Borani jifu wal irra qaba. Tuni, aada nu duraanu jirtu. Yo amma aada bade, ka deefannu jenne, tun waan guddo dansa[. . .].Maalin Adan ka chief *duri abbaan isa Hussein Ido Roble. Ido Roble sun hayyu.*

Ajuran and Boran have obligations to each other. This custom has been there even before us. If now we want to go back to the custom which has got lost, that is very good [. . .]. The former chief Maalin Adan, his father was Hussein Ido Roble. And that Ido Roble was a *hayyu* [a *gada* set speaker, i.e. the holder of an office bestowed by the Boran].

Duri hayyu isa nami lallabe T'uuye Galgalo, adula Warr Jiddaati, akaaku Halake Guyo T'uuye. Gaafasi Borana Ajurani tokko. Maalin Adaniille,

---

[40]To compensate for the under-ageing of their *gada* system the Boran have a parallel age-set system known as *hariya*, into which recruitment is done by actual age rather than generation Legesse (1973), Haberland (1963), Schlee (1989a), Shongolo (1992), Zitelmann (1990) and Baxter (1979).

[41]Schlee (1989a pp. 212f, 219, 227).

Hapiite T'uuyeeti hayyu lallabe. Hapiite Dirre d'ufe, gaari Abdi Fiteetin gambo d'ufaniini Buna geesani.

The one who proclaimed him was T'uuye Galgalo, the adula [seniormost *hayyu*] of the *Worr Jidda* [a clan of the Gona moiety of the Boran], the grandfather of Halake Guyo T'uuye. At that time Boran and Ajuran were one. As to Maalin Adan, it was Hapiite T'uuye who proclaimed him. Hapiite came from Dirre [the former Boran heartland in High Ethiopia], he was brought from the other side in the car of Abdi Fite [a Moyale trader] and taken to Buna.

D'aka dirra Buna korsiise, artiiti booku Ido Roble itti kenne.[. . .] Gaafasiille Isilaanuma ammo aadalle harkuma qaba.[. . .] (Pro) Chaaki! jennaan, harreen gurra d'aabdi jed'ani. Isan amma Jiille qulullu jennaan, ya gurr d'aabne, fiit'om qaabanne.

He took him up the stone outcrop near Buna town and there gave him the sceptre of Ido Roble. Even then the Ajuran were Muslims, they were holding the customs as well. [. . .] 'If you say *chaak*! [a command to drive donkeys] the donkey will raise his ears', it is said. When now you mention Jiile *qulullu* [the bald Jiile, i.e. the members of *Worr Jidda* without a *guutu*, the Boran braid, pointing to the former ritual affiliation of the Ajuran to the Boran] we raise our ears, we remember the relationship.

Dr Guracha, the MP for Moyale then adduced a couple of proverbs[42] in support of the re-discovered Boran-Ajuran relationship.

[. . .] Boni waraan, harka nu qaba. Waraani ka nyaapa harka nu qaba. Tun chuf waan irra malannu. [. . .] Obboleeyan teen ta Ajurana, bonaafi waraanilleen walumaan harka nu qaba. 'Ulma fula takkaati d'ukub' jed'ani. D'ibi keen takka.

Drought is war; it has got us in its grip. The spear of the enemy, too, is pressing on us. Let us consider all that. [. . .] Drought and war are pressing on our Ajuran brothers as well. 'The pains of childbed are in the same place [for all women]' it is said. Our problems are the same.

Nami wanti takkaan harka qabdu, yo mala mari walti naqate guddo dansa. 'Tolaaf hammeenni waliin hin deemu'. Yo finna[43] walti naqanna jenne, ka finnum tola taatu.[. . .] 'Maan horanna, maan soorannaati dura jira'. Akkasiifu nam qara wal arge, odu wal irra fuud'etti, finna duduubatti wal barbaada. Waldeenan arra tun tanum ha taatu.

If people are under the same pressure, it is very good for them to pool their thoughts. 'Gratefulness and evil doings do not match'. If we agree to combine in our management, let that be based on mutual acknowledgement. [. . .] 'What shall we eat?' comes before 'what shall we breed?' Therefore those who have met before and have asked each other for news, can then look for joint way to manage things. May that be the result of our present meeting.

Chief Abdisalaam of Butte, an Ajuran, reciprocated these views:

[42] For the rhetorical and jural use of proverbs see our book *Boran Proverbs in their Cultural Context* (Shongolo and Schlee 2007).

[43] *Finna*, here translated as 'management' stands for the customary way to do things and implies mutual dependence and cooperation.

Afaan keenan, obbolumti walti himan eebti fakkaati. Nuufi Borani arra walti hin d'umne, duru obboleesuma. [. . .] Dira Moyale kana, gaafa jaaran ka ferenjiin d'ufe, nami Ajuran itti matta Mahad Hassan, akaaku aabboti. Gaafas duraati Ajurana Borani walti d'ufe, gaafa gada Abai Baabbo. Gaaf gada tana, Borana Ajurani seer marra bisaani walti keeyate. Gaafas Garrii Dogodin laf tan hin jirtu.

In our language [meaning: culture, customs], to remind one of brother-hood is a shame. We and the Boran have not met for the first time today, we have been brothers before. [. . .] When this town of Moyale was built at the time when the Whites came, the head of the Ajuran was Mahad Hassan, the grandfather of my father. Long before that the Ajuran and Boran have come together, during the *gada* period of Abai Baabo [1667–1674]. During that *gada* period the Boran and Ajuran put up shared regulations about pasture and water. At that time there were no Garre and Degodia in this land.

[. . .] An Boran laf biyya barbaadu takka hin gare. Laf Borana nuhum warra gaalatti barbaada. Lammeesso, nu ya sirkaalatti himne, akka seer keen ka duri Moyaleen walti ejju nuu deebisan, yo isan ka nurra fuutan taate.

I have never seen Boran looking for other people's land. It is only us, the camel people, who look for Boran land. Secondly, we have told the government that they should return our old boundary to Moyale, if you consent to that.

The reader who recalls the slow disintegration of the Boran-Ajuran alliance in the 1920s and 1930s (described in Schlee with Shongolo 2012, Chapter 1 *'Pax Borana'*), the time when the 'Galla-Somali line' was shifted westwards to include the Ajuran on the Somali side, will notice that at this point history has come around the full circle. The representative of the very same area which was then transferred from the colonial Moyale District to what was to become North Eastern Province is demanding that the affiliation to major administrative units is now reversed.

Golich Galgalo, the chairman of the Moyale County Council, then asked a sensitive question resulting from this possible re-affiliation:[44]

Amma finn walti naqan jenne, nu afaan Somali him beennu. Gar keesani, namu afaan Somali dubbata. Yo namummaan d'ibiin, ka akka keesan afaan Somali dubbatu d'ufe, anilleen Ajuran jed'e nu seene, tan dandeetani nuu diddani? Lammeesso, yo finni walti nu dude, ka egeri seer Moyale ka duriiti deebin jettani, lafaaf barchuma keen kajeella qabdu?

If now we combine the management of our lives, we do not know the Somali language. On your side, everyone speaks Somali. If other people who speak Somali like you come and say, 'we are Ajuran as well', and settle among us, can you keep them away from us? Secondly, if now we combine our ways, and if later the boundary of Moyale is put in its old place, will you eye our land and our seats [in political representative bodies]?

---

[44]His real worry might have been that the Ajuran councillors were dissatisfied with what they got as members of the Wajir County Council and therefore wanted to become members of the more affluent Moyale County Council.

The first question, that of people accepted by the Ajuran thereby having access to Boran resources, will have a ring of familiarity to everyone who reads the above mentioned chapter *'Pax Borana'* (Schlee with Shongolo 2012) or the corresponding parts of *Identities on the Move* (Schlee 1989a). The problem is age-old. The chairman alluded also to the Gelible, who still switch between being Ajuran and being Degodia as they have always done[45] as particularly doubtful candidates for acceptance as allies. Because of their situational handling of social identities, the Gelible have already been dubbed 'unidentified survivors'[46] by local English-speaking elites.

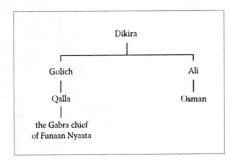

Figure 1.1 The relationship between Osman Ali and the Gabra chief of Funaan Nyaata

Osman Ali, an Ajuran elder, recalled a family relationship to the Gabra:

[. . .] Nu Ajurani, Gabra Boranaan takka. Jiif Gabra ka Funnaan Nyaata, Qalla Golichaati d'ale. Akkaaku Golich Dikira, abbeeruma obbolees kiyyaati d'ale. Garriin duri hammeen Gabra nutti himti, gargari nu him beetu. Gabarti bofa ufiiti hin d'ieesina nuun jetti.

We, the Ajuran, are the same as Gabra, and Boran. The Gabra chief of Funaan Nyaata was begotten by Qalla Golicha. He thus is the grandson of Golich Dikira. He was begotten by my [lineage] brother. The Garre used to tell us about the badness of the Gabra, they did not know about the details. They tell us 'the Gabra are adders, do not let them come close to yourselves!'

Tun chuf yakka, gargari nu fageesiti tan dubbatti. [. . .] Nu yo takka taane, naminu dandae jiddu teen hin seenu. Nu Ajurani, lafaafi barchuma Borana kajeella hin qabnu. Yo ammo inni obboleessa jed'e wa nuu kenne, hinum fud'anna.[. . .]

All this is defamation, by saying so they want to keep us apart. [. . .] If we are united, no one can get between us. We, the Ajuran have no wish to get Boran land or a Boran seat. But if they call us brothers and give us something, we shall take it. [. . .]

---

[45] See Schlee (1989a p. 43, 191) and Schlee with Shongolo (2011 Chapter 1 *'Pax Borana'*).

[46] Note the fine sense of irony in the expression 'unidentified survivors'. After an accident you may have 'survivors' and 'victims' and normally some of the latter, not of the former, may be 'unidentified'. The expression 'unidentified survivors' points to the survival value of ambiguous identifications.

Lammeesso, Borani duri D'adach Warraabiti asi isumaati qubata. Nyaapu-maati kaase armaan d'ufe. Lafti Borani keesa kae sun, haga arraalle maqum Borana qabdi. Nu Ajurani, laf marra bisaani, ta hara eela, maqa Borana suniin falmine Safara Dogodi irra fud'anna. Nutti obboleessatiifi akkan. Gaafasi, afaan Borana ka nu dubbanuufi, falamaan nuu taate.[. . .]

Secondly, the Boran used to settle between here and D'adach Warraabi. Enemies drove them from there and they came here. That land which was left by the Boran, until now has got Boran names [toponyms]. We, the Ajuran, have been claiming the land, pasture and water, pond and well, because of those Boran names and took it from the Somali and the Dego-dia.[47] All this is because we are brothers. At that time the Boran language which we speak, helped us to establish that claim.

Hussein Somo, an Ajuran elder, also had Boran ancestry:

Gosa laf tan qubattu, goci nu afaan wal beennu, ka aada wal beennu, Gabra Boran challa. An Gelbariisa, haati tiyya dubra Dambitu. Abbaan niiti tiyya Warra Jidda Jaaro. An ufi Ajuran afaan tokkocum ka Borana challa beeka. Afaan haad'a tiyyaatifi akkan [. . .]

The tribes which settle in this land, who speak the same language as us, with whose customs we are familiar, are just the Gabra and the Boran.[48] I am Gelbaris,[49] and my mother is a daughter of the clan Dambítu [of the Gona moiety of the Boran]. The father of my wife is of the lineage Jaaro of the Warra Jidda [also Gona]. Myself, I am Ajuran, and the only language I know is this one of the Boran. That is because it is my mother tongue.

To those familiar with descent reckoning in the area, all these relation-ships claimed through the *genitor* rather than the *pater*, i.e. the social father, the one who paid the bridewealth for the mother, come as a surprise. Relationships through women's lovers tend to be systematically ignored in public or official discourse. People might discuss similarities of children in a more or less jocular fashion. No secret is made of known biological relationships. But these are entirely irrelevant for lineage structure, descent reckoning or exogamy. Biological fathers are not real fathers, so one can even marry their daughters by other women than one's mother. That now so many appeals to such relationships are made is an indicator of the intensity of identification work in process. Hussein Somo then continues:

---

[47] Note that the term Somali in the way it is used here does not comprise the Degodia. In the chapter 'Pax Borana' (Schlee with Shongolo 2012) we found a similar distinction between 'Somali' on the one hand and 'Ajuran' and 'Garre' on the other in early colonial records. This speaker seems to reserve the term Somali for more recent arrivals like the Daarood.

[48] This distancing from the Garre goes rather far. Of course, most Garre speak Boran or at least understand it, even if they speak Somali, or Southern Somali dialects like Af Rahanweyn or Af Garrex Kofar as their mother tongue. The linguistic situation of the Garre, however, is complicated. There are Garre who speak English or Swahili to each other because they do not share a single Cushitic language. They are the extreme opposite example to language nationalism: they have a strong sense of political coherence and military interdependence without sharing a language. The Garre, of course, share Proto-Rendille-Somali (PRS) roots with the Ajuran and Islam. Ajuran and Garre culturally have a lot in common. This, however, is implicitly denied by the above speaker.

[49] 'Pax Borana' in Schlee with Shongolo (2012) and Schlee (1989a pp. 159, 212f, 219, 227).   83

[. . .] D'ibi keen guddaan arra, warrum duri nu maqa Isilaanatiin obbolees jenne laf teennatti gadi d'iimnetti, man jaarate, beeshara tolfate, nu kees gabbate, arra duub malfanno, nu had'uuti jira. Isanille tanumaati harka qaba.

Our problems today are big, the people whom we called brothers in the name of Islam and whom we allowed into our land, who built houses, traded, and became fat among us, now we are their easy prey, they are raiding us. They are oppressing you in the same way.

Warri suni dira Wajeera keesa nu baase. Dura godaanne Buna Butte qubanne. Amma armaalle keesa yaa jecuuti jira. Gar itti yaanu d'abnaan, ufirra d'owarre. Tanaafi arra waraani jiddu teen jira.

They have driven us from the town of Wajir. We moved away from them and settled at Buna and Butte. But even from there they tell us to move away. When we missed where to go, we defended ourselves. That is why there is now war between us.

Dr Adullahi Ali, the MP for Wajir North, could not capitalize on any knowledge of the Boran language and had to address the gathering in Swahili. The Somali language would have been too inappropriate on this particular occasion:

[. . .] Jambo la kuleta makabila pamoja sio jambo rahisi. Tumeleta ninyi pamoja ili mkae na kurudisha uhusiano uliopotea kati yenu. [. . .] Ajuran na Gabra Borana sasa tumemaliza mambo yetu, kutoka hapa hadi Marsabit na Isiolo. Hata Orma wataungana na sisi pole pole.

[. . .] The matter of bringing the tribes together is not an easy matter. We have brought you together so that you sit together and bring back the lost relationship. [. . .] We, the Ajuran, Gabra and Boran have now settled our matters, from here to Marsabit and Isiolo. Also the [Tana River] Orma will join us by and by.

[. . .] Sisi hatutaki mambo ya kuzungushana. Kama ni umoja tuwe namna hiyo. Hile style ya Garri, leo pande hii, kesho pande hile, tuwache tabia hiyo. Kama ni uhusiano ya Boran Ali, Dogodi Ali, na baadaye mkasa, hatutaki hivi hivi. Kama sisi Jiille qulullu, tuwe Jiille qulullu kamili bila tashwishi.

[. . .] We do not want things to turn round and round. If it is unity, let us have it. This style [in English] of the Garre, today on this side, tomorrow on the other side, let us do away with that habit. If it is a relationship like that of Boran Ali and Degodia Ali [see above] and then a disaster, that is what we do not want. If we have become Jille qulullu, let us be Jille qulullu entirely and without any doubt.

Dr Guracha Galgalo, the MP for Moyale concluded that like water, which obeys the laws of gravity, the Ajuran were flowing towards the Boran:

Akka Borani baanu, 'Galaani gar lafti laaftufi gul yaa'. Tanaafi amma takka taane.

As the Boran say: 'The river flows to where the land is softer'. That is how we have become united.

To appreciate what it meant that the Ajuran became Jilitu/*Worr Jidda* once more, let us reconsider the earlier alliance between Ajuran and Boran and how it broke up (see Schlee with Shongolo 2012 Chapter 1 '*Pax Borana*').

When the British colonized northern Kenya they perceived the Ajuran as Oromo, at least in their political affiliation. In fact, they had been affiliated to the Boran clan *Worr Jidda* and had given regular ritual tributes to the *qallu* of the Gona moiety of the Boran, himself from *Worr Jidda* as well. Around 1980 middle-aged Ajuran informants would flatly deny that their ancestors have ever been anything but Somali or Arabs and would dissociate themselves from the 'pagan' Boran completely. I have even received such a response from an elderly man whom I knew from the Kenya National Archives to be the holder of a Boran title. By the year 2000 the Kenyan Ajuran managed to re-insert themselves into the Boran fold and much knowledge about their relationship to the Boran a century earlier has surfaced again. This knowledge includes even genealogies through lovers, although there is a strong cultural emphasis about paternity being established by the bridewealth for the mother, and extra-marital relations are known to exist but are normally completely discounted in descent reckoning.

All this knowledge has been kept hidden but alive at some lower level than the official self-description for many decades. It has been kept in store for future eventualities. Our common ideas about oral traditions imply that these are constantly reworked. Unlike written sources, which are subject to social and political forces only up to the point of time when they were written down (although after that they are still subjected to selectiveness like selective attention and selective destruction), oral sources are believed to be constantly reshaped to fit the needs and convictions of whoever re-tells them. This implies that the older versions, which fitted the needs of the past generations, fall into definite oblivion. But is this really so? The case discussed above is about a pendulum movement. The Ajuran once were politically Oromo, then identified as Somali, and recently moved back into the Oromo fold. But the older version of their oral history was not deleted. A backup copy of it was kept somewhere. During decades of denial of ever having had Oromo affiliation, a rather detailed hidden tradition about the Oromo side of Ajuran history must have been preserved by some people.

The phrase 'backup copy' implies a computer image. Problems of collective memory[50] (Halbwachs 1997; Assmann, J. 2000) of the kind we are dealing with here have more often been discussed in connection with conventional, non-digital forms of writing. Writing is certainly a subtle way of forgetting.[51] By committing things to writing, we no longer need

[50] Berliner (2004) criticizes some current uses of the term 'memory' in anthropology. It is not always clear when the psychological process of individual remembering is meant, and when the persistence of cultural items through time is implied. If the latter is the case, what is the difference between 'memory' and 'tradition'? The use of the term 'memory' by anthropologists often is a 'catch all', bordering on the 'sloppy', and 'a haphazard mix of Halbwachs, Bergson, Freud and Connerton' (see Berliner 2004). I find the phrase 'collective memory' quite inspiring. In the present context, dealing with past group identities, the knowledge of which may be preserved or not, we certainly deal with supra-individual phenomena. Also the metaphorical uses of 'memory' (computers, libraries) are good for thinking.

[51] This thought is not new. Assmann, A. (1999 p. 185) cites Plato in this context.

to memorize them. We can then shelve them. We do so with due respect to the fruits of intellectual labour, thinking that by doing so we preserve them. With the continuous production of writings, the accumulation of texts soon fills not only every intellectual worker's private shelves, but also those of the formal institutions of collective memory, the public libraries. Some texts may never be looked at again.[52] To which extent libraries are a store of thought for future use and to which extent they are a graveyard of thoughts which will never be thought again, entirely depends on the unknown acts of future users. But books need to be dusted from time to time; they may fall into one's hands accidentally. Books on a shelf always have the potential of being accidentally rediscovered.[53]

This is in clear contrast to the fate of oral texts. In an oral culture a story which has not been told for a generation dies with the last person who has heard it, and it dies definitely, whereas for the written thoughts in the graveyards of our libraries there is always a chance of resurrection. The Ajuran cannot have accidentally re-discovered their Oromo past after having forgotten about it completely. Also the re-oralisation of written texts can be excluded.[54] We must look for another mechanism for the re-emergence of the earlier tradition.

What Assmann, J. (2000 p. 121) writes about canonization might put us on the right track. With the proliferation of writings in the course of the development of a literate culture, at one point the need is felt to select those texts which are authoritative, which represent the pure religion, the real law, the true history. The texts which become part of the canon preserve reflections of those texts which in the process of selection may

---

[52] There are entire bodies of literature which are materially preserved but which have fallen into oblivion by never being looked at again. Changes in language use may accelerate this process. Eighteenth century writings by Germans in French are an example. Students of German literature read German and students of French focus on France. Well into the nineteenth and early twentieth century doctoral theses were written in Latin. Who has last had a look at one of those theses? There is ample room for re-discoveries.

[53] To remember something implies the re-*presentation* of something which has not been present. If I say 'on my way to the office I remembered that I had to call at the pharmacy', that implies that the errand had not been on my mind and was recalled to my mind. In a way it had been external to my mind and needed to be internalized again. What writing changes in comparison to the absence of writing, is the place of externalization. What is external to my awareness in an exclusively oral culture can be preserved in my long-term memory (the content of which may be beyond the limits of my awareness at a given moment) or in the memories of others (who may re-mind me). Information preserved in a book or in a computer, on the other hand, may be external to my brain and all our brains for a long time, until someone rediscovers it. This is a more distanced form of externality. Only something external to us can confront us, and degrees of externality may correspond to levels of intensity of the feeling of surprise in the case of renewed confrontation.

[54] To the best of my knowledge there is no local Ajuran literature on this point. The use of writing among pastoral Ajuran is very limited. Some elders know the Arabic script, which they use for Arabic only, although it can be as easily adjusted for any language as the Latin alphabet. They use it entirely for ritual and legal purposes. Scholarization of the young is low and so is accordingly the rate of literacy in Swahili and English among them. If the Kenyan National Archives or my own book *Identities on the Move* had been used for the re-discovery of the Oromo past of the Ajuran, they would have been cited, as 'being in a book' adds authority to a statement. Apart from that, the details of the evidence given by elders (above) for the earlier Boran/Ajuran relationship were new to myself and thus cannot be explained by the contamination of oral tradition by my own writings.

be deleted forever, by arguing against them. The defenders of the pure religion have to name the heresies they denounce as such, the systematisers of the law reason against variation and mere local customs and thereby preserve the memory of them, and those who proclaim the true history of a group, which at best can represent only one of many interwoven histories, have to name alternative histories in order to mark them as wrong.

A similar mechanism might be at work in oral tradition. For most Ajuran it might have been sufficient to know the 'official' (i.e. the publicly proclaimed, positively evaluated) version of Ajuran history, namely that the Ajuran are Somali of ultimately Arab descent and have been Muslims from the early days of Islam. Leading people, people whose utterances mattered, people in positions to play identity politics, in addition to this official history might have been taught what to deny. Is this the way in which the memory of the Oromo past of the Ajuran has been preserved? Has this memory profited from the paradox of censorship – a paradox which results from the fact that, in order not to say what falls under censorship, one must have a precise idea of what one is not allowed to say?

While the Ajuran found their way back into the *Worr Libin* alliance under the umbrella of the Boran, the Garre moved the opposite way. They identified themselves more clearly as 'Somali' than at any time before. Being Somali was equated with being Muslim, and their political rhetoric was that of *jihaad*. During a demonstration in Moyale, Garre youths and women sang:

> *Olkiin arra jihaad. Nami kufaar tokko ijjeese reer jannaada.*
>
> The war of today is *jihaad*. Whoever kills one infidel will belong to the people of heaven.

The following text is from an audio cassette, produced in Ethiopia and mainly referring to Ethiopian events, which sold for 200 Ksh in many places in northern Kenya and Ethiopia in April, May and June 2000. It was also available in Nairobi. They were advertised by word of mouth and at a later stage distributed in the small kiosks which also sell *miraa* (*qat*) and cigarettes. The A side of the cassette contains Islamic instructions about the *ahaadith*. Only the B side is political. As it was quite reasonably thought that the content might be offensive to Boran, at first this cassette only circulated among Garre. Adversaries may be or may become similar to each other. Stylistic similarities with the Boran poet Jarso Waaqo Qoto cannot be overlooked (for samples of his poetry see Schlee and Shongolo 1996; Shongolo 1996).

> *Bismillahi Rahmani Rahim[. . .]. Amma waan itti bilaaw waan gaaf jahada yaani waan gaafa Garriin huurat gadi seente. Ka isiin gadi dagaalamte alanki barbaanna jette gad kaate. Gaafa isiin yaani maqaan Islamadi ka gubbaati hafu ka gaaladi nu jal maru jette oli gadi kaate. Duuba tariqi isiin jal qabateefi haga isiin muumme itti baate isanii bilaawwa.*

In the name of God, the Benevolent, the Merciful [. . . in Arabic; continuation in Boran:] Now, what I start with, is about the time of *jihaad*, when the Garre went into the bush. When they mobilized, when they said 'we seek nationhood [literally: a sign, a heraldic sign, a flag]', when they rose. When they said 'May the name of Islam stay on top and that of the infidels come under it', as they rose. I now start with the history from the beginning to the end.

Bismillahi Rahmani Rahim. Ganna kum tokkoofi d'ibba sagali jaatami sadi jia jaa, tariqi kud'ani sadi, guyya sun hid'anne jahada kaaneni qawwe alaman jed'ani, kud'ani shaniin. Gaaf jahadan kaan suni, nam digdami lamaan baane, haga ganna afurtama hafu dinne baane, hafuursine adowgi faan yaane.

In the name of God, the Benevolent, the Merciful. In the year one-thousand and nine-hundred and sixty-three, the sixth month, the thirteenth, that day we girded ourselves and went on *jihaad* with fifteen guns called *alaman* [*almaan* = German guns?]. At that time of *jihaad*, we went with twenty-two people, for forty years we have been fighting continuously, we pursued the enemy openly.

Tariqi Somali Abbo tun akkanaafu yo faan d'awani fago. Barbaad balcum tana nu hin ramne eegi yoomi. Gaaf inni nuun kae jeneraliin keen Aliow. Balcum nu qaabsise nuhu gaafasi daalle gaala looni.

The history of the Somali Abbo, if you follow it, is long. Seeking freedom we have not slept ever since. When it started, our general was Aliow [Gabaaba, brother of Chief Hassan Gabaaba]. He reminded us about freedom, we who were the fools of the camels and the cattle.

Wanti sun isa qaabsiseellen	That which reminded him of that
d'iiratti chidi nu nyaacise	were men who made us eat husks
udaan harre nu nyaacise	they made us eat donkey dung
chid'aan nurra chire	they cut our penises
akka uwwoolee nu yaasise	they made us live like women
biyye haad'a teennalle teesum nurra bite	they forced us to pay for the right to stay in our motherland
gibir huduu nu baasise	they made us pay taxes for the right to sit down
chiniinsun garbumma	the pain of vassaldom
ka chiisu d'owwe nu kaasise	which made us not to lie but to rise
injiraan nu nyaacise	it made us eat lice
injijji mataati nu baasise	they made us get lice eggs on our heads
Garbummaati nu gadi chone	Vassaldom oppressed us
foon lafe irra nu chore	and made us slim down to the bones
chiisu dinne kaane	we refused to lie down and rose
teen balcumma la butanne	we grabbed our freedom
akka qore	like a kite
waan amma gara nu fayyisan	what now will heal our bellies

balcum teen nuu bald'isa	make our freedom wider [more widely known]?
D'iiro kambuni Abdulkadir kuni	Men, the company of that Abdulqadir[55]
gar Oromo gul nu hirkisa	draws us towards the Oromo
hookamo Sheikh Adanti nuu kenna	let us give the power to Sheikh Adan
me gar Somali gul nu gadi fachisa	he will steer us towards the Somali
nami him beekne hin jiru	there is nobody who does not know this
ballaallen.[. . .]	even the blind ones.[. . .]
Akka Garri Rabbiin ol deemsisu	So that God may raise the Garre
hagi ulamaa chufti ha eebisu	let all the learned ones pray for them
Amin jed'a, me haga Rabbiin	by saying Amen, let God make
Sheikh Adan durum keen deemsisu.	Sheikh Adan walk ahead of us.
Ir jirti teesani, Garri him beetani	What is good for you, you Garre do not know
marchuma keesaniille yo fuutani	even if you take your seat,
waggaan hin geetani	you do not keep it for a year
ganna gadi maruufi haaraum feetani	every year which comes around you want a new one[56]
kaan hama jettani biruma deemtani	you say 'it is bad' and leave it
ado sila tokkum challa irr teetani	if you had sat just on one
yoowanna ya fagoon uf geetani	by now you would have reached far
isan akkamiin ur geetani Garri?	how then can you reach anywhere, Garre?
Qiimma teesan tanaan qottani laf jal balleesitani	You dig and remove the earth from underneath your standing
tariqi abbooti teenna ha qaabannu	let us remember the history of our fathers
me, ta barana nu buutelle irra wa ha barannu	let us also learn from what happened to us recently
womu nuu hin baane	it helped us in no way
qawween nu barana mata saabannu	the guns on which we used to rest our heads

[55] As mentioned in Schlee with Shongolo (2012), in the 'The Impact of War on Ethnic and Religious Identification in Southern Ethiopia in the Early 1990s', the original Somali Abbo Liberation Front was renamed Oromo Abbo Liberation Front. The OALF later split, and a part of it renamed itself SALF. Also other movements came up which identified with the Somali cause, like the Somali League Party and the Ogadeen Liberation Front. Abdulqadir 'Issa, a regional representative and the vice-chairman of the OALF, kept the label 'Oromo' for his faction, probably with the Gabra electorate within Region 5 in mind.

[56] The Ogadeen Daarood and the Marexan Daarood as well as the Degodia had politically established themselves earlier in Region 5 and occupied many seats in the regional parliament, while the Garre, who were latecomers to this political process because they had been oscillating between the Oromo and the Somali affiliation, were restricted to lower levels. Many Garre tried to change constituencies in order to acquire representation at higher levels.

womu nuu hin baane	served us nothing
gumaiin rasasa ka nu beesen barbaadannu	the boxes [?] of bullets which we sought to buy
womu nuu hun baane.	served us nothing.
Qabilaan nu gargari baafannu	The clans which we sort apart
ado nu qabila gargari baafannu	as we are sorting out clans
gumai beesen barbaadannu	as we try to sell single bullets
kalaashi keen mata saabannu	and rest our heads on our Kalashnikovs
adowgin nurra faite	the enemies carried them away from us in loads
qaalufi qaro teen fitt'e	They killed our sheikhs and knowledgeable people
aduunya teen lafa tiirte	they swept our property from the earth
qutum keenna gargari hiite	they dispersed our settlements
gurki Garrii hin taane	the marriage of the Garre has not taken place
isanu hin beetani gaafasi	you know that when
qutumi Lae kuni	there at the settlement of Lae
ka idinki bisaan d'ibe	when our fighters had a water problem
rasaasalle shaain irra bite	and a cup of tea was bought for a bullet
ka hin qabneelle beelan fit'e	and those who had none suffered hunger
garaaca Garrii gargar hid'e	the feelings of the Garre were divided
sun adowgumaa kara hiike.	that opened the ways for the enemies.
[...]	[...]
gaalki at waan siin tissuuti jiran	Infidels, we are looking out for you,
bireeni chufa waliin huursu	all the Bren guns will cough together
waan sii gadi tirsuuti jiran	what is being dragged towards you
qalofti chuf waliin yuusu	all the bullets will yell together
waan sii gadi tirsuuti jiran baroolen	what is being dragged towards you
fagooti d'ufe keesa si guru	the portable missiles will be fired
ka mata ke barbaade	from afar and remove you
reef ke wal irraati tuulu	what is meant to provoke your anger
	will pile up your corpses

90   *[...]*   [...]

khalbi hammeen khabduun,	with badness at their hearts
sii chichiisutti jirani	they lie waiting for you
talidi tiiro khabduun, sii gadi	with the commitment and the support
shiishutti jirani	they have, they point the gun at you
daandi Ballalle baate badda	the path which leads from Ballalle to the
Magaada d'akhtu, sii gadi chiruuti jiran.	highland of Magaada is being cleared for you.
Obboleeyan tiyya	My brothers,
diqqaa guddaan nu kaasa	young and old, let us rise
jahaada raasatti nu baasa,	let us go to the bush for *jihaad*
waan harka qabdaniin	with what you have in your hands
gaalki onne isa tarsaasa,	cut the hearts of the infidels to pieces
haalo aanan gaal keenna	the vengeance of the milk of our camels
maati isa marsaa hobbaasa,	round up his family and finish them
gaalki d'iig isa lolaasaa	make the blood of the infidels flow
allaatifi waraabessa nyaacisa,	make vultures and hyenas eat them
haga badda Magaadalle	up to the highland of Magaada
oriisumaan gul nu yaasa,	let us chase them in one go
chiniinsatti chiniinsa sirra bukhisa	pain is driven out by pain
hand'uraati duuda dubbisa,	the legal property makes the dumb speak
nu akkamiin teenne laalla	how can we just sit and watch
ka amma gaalkin aanan gaal keenna	the infidels lie down satiated with the milk of our camels
chiise ilmaanin guddisu,	and raise their children on it
gaalki haati badde galat Rabbi keenna.	thanks to the Lord, the mothers of the infidels are lost.
Arra Garrii la beekatte	Today the Garre have become aware
la gadi baafate	they have come out
bineeyin arra haalo teen baasitu	the wild ones who will carry out our vengeance today
ta d'iig kanke lolaasitu	who will make your blood flow
ta jagna ke kolaasitu	who will castrate your brave ones
ta Raaba gada atin abuudulle	who will make the *raaba gada* you believe in,
finchaan harree obaasitu	drink the urine of donkeys

91

ta qutum ke roraasitu	who will shake your settlements
ta haate lafa si baasitu	who will attack you and chase you out of the land
ta guulki Garrii ol aansitu.	those who make the victorious Garre rise to the top.
Si'i farsoofi bakhti nyaatu	You who drink alcohol and eat unclean
ka boqorkiin ke waraabes waliin	flesh whose leaders compete with
bakhti saamu	hyenas over carcasses
badiiti si tolce male	it is because you are going to perish
ati guulkii akkan hin taatu	you are not doing so for victory
wakhtaati d'io jira	the time is close by
ka geesigiin keen d'aadatu	when our skilled ones will proclaim
ka raabi ke guutu haadatu	that your *raaba* shall shave their tresses
ka Abbaan Gadaallen ashaadatu.	off and the Abba Gada will pronounce the *shahaada* [the two articles of the Islamic faith, i.e. convert to Islam].
Sabbo sadeen ijjeesi	Kill the three Sabbo
akk injiraani	like lice
Gona torbaan ha goolani	let the seven Gona be burned
akka qoraani	like firewood
mirga bita uf laali	look out right and left
d'iig ke ulaan barbaadi	for a place for your blood
d'iirti si oofutti jirtu tun	these men who are driving you now
ta duru ati beetu taani	are the ones you have known for long
ta gaalki bombeesitu	who make ashes of the infidels
ta Dargi balleesite taani	the ones who died away with the Derg
ta dowlan chuf beeku taani	who are known to all nations
ta Gooron bureecise taani	the ones whom Gooro made parade
ta Raabi him beene taani	the ones whom *raaba* do not recognize
ta diina gul deemtu taani	the ones who follow the religion
ta fulaan wa geetu taani	the ones who make something succeed
Gardallo martitti	insolent foolish people
akka duul allaatii	like a swarm of vultures
waan guddo sii taate	did a lot for you in the past
duri Mangistuufi Amarti	in the time of Mengistu and the Amhara

fula warri keesa kae	in the place these people left
at teete barbaada	you want to establish yourselves
ibidaati d'umatte	you will perish in the fire
imiimi rooba taata.	you will become like the tiny insects which come out after the rain.
Gardallo daalle	Insolent foolish people
akhli malaasa	with brains full of pus
me uf laallad'i	mind yourselves
akhli te doomaa	your brains are hornless
eesatti d'eebu baata	where will you quench your thirst
guyya jaba bona	at the time of a severe dry season
tulla saglaan	the nine great wells
amma d'iiratti tooa.	now men draw water from them.
Dagaalkin waan jiran	The struggle has been there
saban Nabiyaashi keenna	since the era of our Prophets
yo dagaalkii doonelle	even if we die in this struggle
qubqabaadd'a qar reer jannaada	know that our souls are already among
nafseen teen.	the people of paradise.
Soomali gaal keen	Yellowish are our camels
Somaliin gos teen	Somali is our tribe
Hassan maqa keen	Hassan is our name
ha nuu taatu	let our repentance
tooban teen.	be accepted.

This text, which has the form of a sermon and the content of a harangue, equates the struggle of the Garre with a *jihaad*. A closer look however, reveals that not only are Islam and Garre nationalism mixed, but that many elements of the Cushitic killer complex creep in (see Schlee 1989a, Schlee with Shongolo 2012). Islam explicitly forbids the mutilation of slain enemies, a practice which offended the Prophet at the battle of Uhud. The threat of castration, which is an allusion to the practice of taking genital trophies, is definitely un-Islamic. There are other observations which contradict the simple formula Boran = infidels, Garre = Somali = Muslims.

In another volume (Schlee with Shongolo 2012) we have a chapter on the impact of war on ethnic and religious identification which refers to events in the early 1990s. There we have noted that many Boran withdrew from mosques whenever their community was at war with Somali. The Garre then as now referred to the Boran as *kufaar*. Then as now this derogatory designation was applied, although some of the Boran were and are Muslims.[57] In the early 1990s, however, the Boran mirrored the Garre

---

[57] For a long time the Muslims of Moyale have been divided along ethnic lines. Somali frequent the Masjid Mur (mosque), the Boran the Jamia Mosque in the centre of town. Also the minor mosques at the outskirts reflect the ethnic composition of the neighbourhoods.

view of them by primarily identifying Islam with being Somali. Under attack by Somali groups the non-Muslim Boran and those who were only superficially Islamized would distance themselves from Islam and criticize practicing Muslims from their ranks as following Somali customs. In the recent clashes they did not do so. Islam was no longer regarded as a Somali attribute. Instead even in this period many Boran were converted to Islam as far inland as Dubuluq (in the central Boran region of Ethiopia). With the pan-Boran solidarity strengthened by the Garre threat, Muslim Boran scholars from the Waso area (Isiolo District) preached and proselytized all over Boranland.

One such Muslim scholar, a Boran from Isiolo, Sheikh Abdullahi Golicha, was in the area preaching at the time of the clashes. The Garre community of Moyale who declined even to pray behind a Gabra Imam in one of their mosques, invited him to preach to them. In his sermon, which was partly in Swahili, he reminded them that it was in the month of Muharram, during which waging war is forbidden to Muslims.

> [. . .] *Ninyi mnawaita WaBorana na WaGabra ati wao ni kufaar. Badala ya kuleta hawa kwa dini ya Isilamu vile ume amuriwa na Mungu, mnawatorosha. Mwezi huu sio mwezi wa vita masheikh wenu wanajua hivyo. Hata katika aada ya WaBorana, mwezi huu WaBorana wenyewe hawaendi kupigana na maadui.*

> You call the Boran and the Gabra *kufaar*. Instead of bringing them to the religion of Islam as you have been ordered by God, you chase them away. This month is not a month for war, and your sheikhs know that. Also in the Boran customs, the Boran themselves do not go to fight enemies in this month.

On several occasions, Garre sheikhs joined him in preaching about Muslim brotherhood and about restoration of peace to Moyale. Councillor Edin Adow of Godoma location, one of the alleged Garre warlords 'wanted' by the Boran, attended an evening prayer at Butiye, a predominantly Boran settlement, where Sheikh Abdullahi had come to preach. No one questioned him on this occasion.

After considering the Garre/Boran relationship it is worthwhile to look more specifically at the Garre/Gabra relationships in the light of the recent changes of alliance. Gabra Miigo, i.e. the northern Gabra to whom our sources refer in this context, have largely the same clans as the Garre. They share the Proto-Rendille-Somali (PRS) origin. Both groups now are Muslims, although the Gabra are said to be so only in town. The ethnic split between them occurred when, centuries ago, the Garre managed to withdraw from Boran dominance for a while to the east and the Gabra submitted to it with less delay. This is what, in modern parlance, makes the Garre more 'Somali' – although in 1990 Schlee spoke to many Garre who still rejected that label – than the Gabra. The latter have also kept their peculiar form of the *gada* system (Schlee 1989a pp. 82f) and, in connection with it, the ritual journey to their holy mountain Hees, on the top of which a camel has to be sacrificed, as distinguishing features.[58] We have described elsewhere (Schlee with Shongolo 2012) that in the early 1990s there was a split between the Boran and the *Worr Dasse*, 'the people of the mats', the camel keepers who were the former allies of the Boran. At that time, the Gabra and Garre found themselves on the same side of the dividing line. Thus in 1992, during the conflict between the Boran and the Gabra/Garre, the Degodia had the opportunity to occupy the niche left by the two ethnic groups in relation to the Boran. The new Boran/Degodia alliance still worsened the rift between the Gabra and the Boran. As time went by, the Gabra realized that they had made a mistake by forming an alliance with the Garre who did not allow them access to political representation. They did not allow them access to better pasture and water either. The Gabra found themselves in the same predicament as earlier, in their decaying relationship with the Boran. 'They jumped from the frying pan into the fire or they jumped from the python into the cobra's mouth' as one elder expressed it. There was no way for the Gabra to return immediately into the Boran fold. An opportunity came when the relationship between the Boran and the Degodia deteriorated. To this, the Gabra might have contributed. The Degodia thus in 1998 became a common enemy both to the Boran and the Gabra. Without considering their past differences, the Gabra at once renewed their alliance with the Boran to fight against the Degodia. As the Ajuran, who returned into the Boran fold as well, were at war with the Garre, the Gabra now found themselves in alliance with the enemies of their prior allies, the Garre. In a 2002 draft of this chapter we concluded this paragraph with the sentence: 'We now have Boran/Ajuran/Gabra against Garre and Degodia, the latter two also at odds with each other. The Gabra did not like the Ajuran to occupy their old position closest to the Boran. Also their former ecological niche on the outskirts of Boran grazing lands was now occupied by Ajuran. It remains to be seen how stable this alliance will be.' This scepticism in the meantime has become more than justified. Further down we will have to discuss the complete deterioration of Gabra/Boran relations. But in our chronological account we now return to the year 2000.

In addition to local causes, it can be said that escalations in one place stimulate those in another. The clashes in Isiolo took place in the month following the Moyale events. Boran politicians from Moyale, Marsabit and Isiolo share the same road to Nairobi. They often travel together and appear together in public. They visualize and articulate the common Boran cause. The Isiolo events and various comments on them from different political perspectives are summarized rather well by Konchora Guracha in the *Daily Nation*, 20 May 2000 (p. 5):

---

[58] On the holy mountains of the other branch of the Gabra, the Gabra Malbe, and the age-set promotion rituals associated with them, see (Schlee 1990a, 1992d, 1998b). These holy mountains have great significance as symbols of social identities. The rituals performed there and their timing further illustrate the interconnectedness of structurally quite dissimilar *gada* systems through the transfer of ritual services. The Garre, by the way, although they do not have a *gada* system of their own, are not outside this network (Schlee with Shongolo 2012).

'Land and pasture at the heart of the killings in Isiolo'
'Conflict is likely to go on'

It is Saturday morning in Isiolo town. Small groups huddle in discussion. They are trying to come to terms with a thunderbolt: the sacking of their local member of parliament, Mr Charfano Guyo Mokku [as an Assistant Minister] the previous day.

There are rumours that the crowd is contemplating a major demonstration to protest at their MP's sacking. We defy local counsel and drive to the killing field in Isiolo. [. . .]

We set off for the town again to record the planned demonstration. As we near Safi Estate, we see vehicles speeding away. Fleeing crowds follow them on foot and herds of camels. Trouble broke out at the demonstration as people threw stones and burnt houses and commercial buildings.

Welcome to lawless Isiolo District, independent Kenya's latest dark reality, where an estimated 100 people have been killed in fighting between Borana and Somalis.

Two weeks ago, some 40 people, among them three policemen were killed in the Emiret area. The Degodia have been accused of being the cause of the recent killings.

In 1991, some 3,000 Degodia pastoralists crossed into the Merti Division, inhabited by the Borana, with their livestock in search of pasture and water. The agreement with their Borana hosts was that they return to their communal lands in the neighbouring Wajir District in North Eastern Province.

The agreement was not honoured and tension started building up. The situation became particularly volatile as the 1997 General Election approached. Even the government was cautious about enforcing the relocation of the Degodia to Wajir.

Local politics complicated the matter further as politicians saw a chance to cash in on extra votes. Eligible Degodias are allowed to register as voters and participated in the election.

The pastoralists were eventually pushed out of Merti Division and settled in the Kipsin area, where they have fought bloody battles with the Samburus from the adjacent Ol Donyiro grazing lands.

Insecurity has taken its toll on the lucrative tourism industry, on which the Isiolo County Council depends. It earns more than Sh 80 million from the industry. The district hosts popular tourist facilities like the Samburu Serena, Sarova Shaba, Samburu Intrepid and Bufallo [sic] Springs.

The council uses Sh 10 million every year for bursaries for needy students and recruits and train its own rangers. Somalis have accused the civic body of discriminating against them in employment. [. . .]

Another contentious issue in the Borana/Somali feuds is the argument that the Somalis were endangering the environment with their large herds.

'The large number of herders around the parks and reserves and the prevailing insecurity is retarding our community conservation efforts and development', said Mr Dabasso Halkano of the Detha Forum, a local conservation body.

Land is also at the heart of the Isiolo clashes. According to the archaic Trust Land Act, all pastoral land is held in trust for the people by the government. Recent cases of illegal land allocations has [sic] not earned the government the people's trust and security of tenure is difficult to ensure, partly due to the nomadic nature of the pastoralist. The invasion of the Somalis has not been taken kindly by the Borana, to whom land has a near-sacred significance.

The deployment of the Army to control the situation was controversial as many complained that the soldiers used excessive force and an estimated 100 people were killed.

The sacking of the assistant minister, a Borana, did not help ease the tension. His people perceived it as persecution of their MP, who was fighting for their rights.

Leaders had different views about the cause of the hostilities. Wajir West MP Adan Keynan (Safina)[59], a Somali from the Degodia clan, said he could not rule out local political involvement in the killings.

Mr Mokku claimed that 'powerful external forces' were involved. 'The control and regulation of the movement of livestock, as was done in the colonial period, has completely broken down. And unless the government moves in to stop the influx of firearms, there will be continued lawlessness', said Isiolo South MP Abdullahi Wako.

'Political leaders are not sincere. Clashes over pasture and water were often spontaneous and not pre-meditated. This (Isiolo) is political', said former Lagdera[60] MP and now Safina chairman Farah Maalim.

'This is political banditry where some rich tycoons use their wealth to set people against one another. Provisions of the Trust Land Act must be strictly enforced. The Constitution must be enforced so that law and order prevails', added Moyale MP Guracha Galgallo.

'Arms are widespread in most of northern Kenya and politicians are not to blame for the violence', said Fafi MP Elias Bare Shill (Safina).

'The Boran and the Issaks [Is.haaq, Isxaaq, Isaaq] have lived together peacefully since 1946. The current pressure to evict Somalis has a political angle to it', claimed Kimilili[61] MP Mukhisa Kituyi (Ford Kenya)[62].

For Ugenya's[63] James Orengo (Ford Kenya), the Isiolo ethnic killings are just one case of 'a policy the government has perfected to intimidate opponents'.

The people in the eye of the storm – the area residents – seemed to have a rather practical view of the problem.

Said one man fleeing atop a land Rover at Safi Estate: 'All I know is that the Borans and the Somalis are fighting.'

Following the bloody clashes between Boran and 'Somali' groups described above, Isiolo North MP Charfanna Mokku was sacked as Assistant Minister for Information, Transport and Communications. On 11 May 2000, C. G. Mokku went to Isiolo with three other members of Parliament, Hon. Dr Guracha Galgalo, MP for Moyale and Assistant Minister for Health who has been extensively quoted above, Hon. Abdi Taari, MP for Saaku (Marsabit Mountain) and Assistant Minister for Energy, and Hon. Dr Abdullahi Waqo, MP for Isiolo South, another Assistant Minister for Health.

About 2,000 people were waiting to welcome them eight kilometres outside Isiolo town. They were singing victory songs, *gerars*, and chanting slogans condemning the Government's action to sack their MP, Charfanna

[59] Safina, an opposition party, was founded by Richard Leakey and others in 1995.

[60] Lagdera is a constituency in larger Garissa County.

[61] Kimili is a town in a rich maize-producing area in Bungoma County in Western Province.

[62] Ford stands for Forum for the Restoration of Democracy. It was founded by Oginga Odinga. In 1992 it split to become Ford Kenya and Ford Asili, led by Kenneth Matiba. In 1997 there was another split: Raila Odinga, a son of former Vice President Oginga Odinga, with a number of other members, left it to join the National Development Party.

[63] Ugenya is a town in Siaya county of Nyanza province.

Mokku. One placard read, 'Welcome to Isiolo. Isiolo was the district of peace but now of bloodshed and Opposition. Let the intruders return to their homelands'. Another placard, in Boran, read: *Warri miilan d'uftuun ka lafa teen keesa galu* – 'Those who came by foot should leave our land'.

The cars bringing the politicians were greeted with cheers and ululations. The MPs were carried shoulder-high into town. Then the chairman of the Isiolo County Council addressed the crowd in Swahili:[64]

> *[...] Vita ambayo vinaendelea katika district hivi vitatu ya Borana, Isiolo, Marsabit na Moyale, ni vita ambaye imeletwa na watu wa majirani ya North Eastern [...] Tumesema mara kwa mara tangu vita vianze hapa Isiolo ya kuwa sisi jamii ya Waborana hatuna shida lolote na jumla ya Wasomali bali tuna shida na wale wafugaaji ambayo wali toka Mandera, Wajir na Garissa districts, na sasa wanaishi katika madistrict yetu ya Isiolo, Marsabit na Moyale. Tumekaribisha hawa wakati wa shida ya jilaali na sasa wamekataa kurudi kwao na wameanza kutuchokoza na kuleta maafa katika wilaya zetu. Na wameanza madai ya ardhi zetu.[...]*

The war which has been going on in these three districts of the Boran, Isiolo, Marsabit, and Moyale, is a war which has been brought by our neighbours from North Eastern [Province . . .]. We have said again and again since the war has started here in Isiolo that we have no problem with the Somali community as such but we have problems with the pastoralists who come from Mandera, Wajir and Garissa districts and now live in our districts of Isiolo, Marsabit and Moyale. We welcomed them in the time of distress of the drought and now they have refused to go back home and have started to provoke us and to bring death to our district.

The chairman then handed over to Charfanna Mokku who introduced the other visitors. Dr Guracha Galgalo, the MP for Moyale, then spoke, also in Swahili:

> *[...] Hile shida ambayo ilituleta hapa ni shida ya madai ya ardhi yetu. Hatujakuja kwa sababu Hon. Mokku. Mokku amenyaganywa cheo cha Assistant Minister. Hiyo ni kitu kidogo. Hajanyaganywa cheo hicho sababu ya kuiba mali ya serikali kama wengine. Amenyaganywa kwa sababu amesimama imaara na watu ambayo anawawakilisha, watu wake ambaye wamedulumiwa. Jambo la vita vya kabila katika Kenya mzima halikuletwa na Muheshimiwa Mokku. Vita hivi vilikuwa vikiendelea kabla ya Mokku hatakuzaliwa.[...]*

The problem which has brought us here is the problem of the claim to our land. We have not come because Hon. Mokku was removed from the rank of Assistant Minister. That is a small thing. He was not demoted from that rank because he stole Government money as others did. He was demoted because he stood firm with the people he represents, his people who were robbed. The matter with the war between tribes in all of Kenya was not brought by Hon. Mokku. These wars were going on even before he was born. [...]

> *Kuna watu ambayo wanaeneza uvumi ya kuwa Muheshimiwa Mokku ndio alisababisha vita katika Isiolo. Ninawauliza hivi: Kuna vita vinaendelea*

---

[64] The event was videotaped. Copies of this tape were sold in Nairobi for 1,000 Ksh. The following transcriptions are from one of these copies.

kule Moyale, Wajir, Mandera na Garisa. Vita hivi ni Mokku pia alisaba-bisha?

There are people who are spreading rumours that Hon. Mokku was the one who caused the war in Isiolo. I ask them: There are wars going on in Moyale, Wajir, Mandera, and Garissa. Was it Mokku who has caused also these wars?

Shouts from the public: *hapana, ni uwongo! –* no, that is a lie!

[. . .] Waacha niwaeleze sababu ambao imeleta shida hapa Isiolo. Hile shida ambaye inatukabili Moyale pia. Kuna wajamii ambayo wameleta mifugo kutoka North Eastern Province. Wameingia kwetu. Tumewapatia malisho. Tunawapatia karibu kila kitu. Na siku ile wanaamua kuondoka, wanaua watu wetu, na wanaondoka. Na sisi tunawauliza kila wakati. Ni kwa sababu gani kama tumekukaribisha nyumbani, tumekusaidia, munatu-fanyia hila?

Let me tell them the cause of the troubles here in Isiolo. It is the problem which we encounter in Moyale as well. There are communities which bring their livestock from North Eastern Province. They have come to us. We gave them pasture. We gave them about everything. And the day they decided to leave they kill our people and leave. And we ask them every time. What is the reason, if we have welcomed you at home, if we have helped you, why do you then do evil to us?

[Switch to Boran]

Ojja lafti nama abaarti taate ganna chufa, abaarti isanti lafaati buuse? Ojja lafti warra abaarti taate, yo laf teen d'ufan ka naga nuu kennani, ka aada teenna heshimani.[. . .]

If there is a drought in the land of those people every year, is it you who brought the drought? If there is drought in the land of those people and they come to our land, let them give us peace, let them respect our customs.

Maqa tokkoon nuhu viongozi teesani d'io tana nuun kokisani. Ojja akkan wa afaani nu bahaniiyu, 'oo OLFiiti keesa dubbata', jed'ani. Isan nami takka OLF arma keesatti arge jira? (..hiyo, hiyo). Waan nu warri lafa, ka Moyale, ka Isiolo, ka Saaku hin argini, guyya chufa OLFiiti warra keesan kees jira jed'ani, ka abba irra beeku maani? (buda! buda!) Budaati nu nyaate yoosi!. (...d'uga, d'uga!).

We, your leaders, are threatened of recent by one name. Whenever anything comes out of our mouths, they say, 'oh, the OLF is speaking inside them'. Is there anyone among you who has seen the OLF here? [Shouts from the crowd: no, no.] What we, the people of the Moyale, Isiolo, Marsabit area have not seen, they say every day 'the OLF is among you', who should know better than we ourselves? [literally, citing a proverb: who knows better than the father/owner?] [Shouts: the evil eye, the evil eye]. Then it is the evil eye which is consuming us! [Shouts: true, true!]

[. . .] Gaafa baran Moyaleeti d'ibi gale, an biy teen akkan gaafadd'e.' Lafti Kenyaa, ta nam chufaati abba hin qabdu jed'ani, beeni sirbe, Moyale abba hin qabdu jed'ani'. Isaniille arra hin gaafadd'a: Central Province ta eennu?

When recently the problems came to Moyale, I asked our people like this: 'the country of Kenya, they say it belongs to everyone, it has no owner, 99 they sing, they say Moyale has not got an owner.' I also ask you today: to whom does Central Province belong?

[Answer from the crowd:] *ta Kukuyuu.*

To the Kikuyu.

Machakosi ta eennu?

To whom does Machakos belong?

. . . ta Wakambaa

To the Kamba.

Meeron ta eennu?

To whom does Meru belong?

. . . ta Meero.

To the Meru.

Sareele woma hin taatin, lafti abba hin qabne, akkamiin teesan challa taati?

You worthless dogs, how is it that yours is the only land which has not got owners?

(Applause and prolonged frenzied shouts from the crowd. Women are crying.)

Later the Member of Parliament for Isiolo South, Dr Abdullahi Waqo, addressed the crowd. He described the 'shape' of the Kenyan Oromo community, by pointing out the distribution of the Oromo language (failing, however, to note that large segments of the Garre and Ajuran, too, speak Oromo):

[. . .] Kutoka Moyale, Marsabit, Isiolo hadi Tana River afaan tokko dubanna-ka afaan Borana jed'ani. Nuhu chufa waan takkaati harka nu qaba. Arra wanni nu wal qabanne d'umneefi, warra Nairobiiti, Eastleigh keesa chiise Isioloon abba hin qabdu, cosmopolitani jed'u. Isioloon ta Borana jenne, itti himu d'umne.

From Moyale, Marsabit, Isiolo to the Tana River we speak one language – it is called the Boran language. We all are in the grip of the same thing. The reason why we today have all come together, is that people in Nairobi who hang about in Eastleigh, say that Isiolo has no owners, that it is cosmopolitan. We have come to tell them that Isiolo belongs to the Boran.

Finally the man whose demotion as an Assistant Minister had caused the whole gathering, Charfanna Mokku, MP for Isiolo North, addressed the crowd in Swahili:

[. . .] Jamii wote wanaoishi hapa Isiolo wanifahamu vizuri. [. . .]. Kuna watu wanasema kule Nairobi ya kuwa Isiolo ni lease area ama trustland, haina mwenyewe. Ni Wazungu wamepeana jamii fulani nineteen thirty-five. [. . .]. Katika Isiolo North, kuna makabila forty-two, lakini majority ni Waborana.

100

[. . .] All communities living here at Isiolo understand me well. [. . .] There are people in Nairobi who say that Isiolo is a lease area or trustland, that it has no owner. It is Europeans who gave it to a certain tribe in 1935. [. . .] In Isiolo North, there are forty-two tribes, but the majority are Boran.

The colonial policies involving tribal land rights in Isiolo District to which the MP Mokku here alludes are clarified in a letter to the editor of the *East African Standard*, by a certain Ismail Sheikh, Nairobi, found in the issue of 10 May 2000:

> Statements made by Borana politicians amaze me and one made by Minister Marsden Madoka on BBC radio on May 2 that Borans are 'Owners' of Isiolo.[. . .] The district is divided in two parts which is well elucidated by a letter written in 1955 to Mr R. Thompson, the then Isiolo District Commissioner. The letter with a reference No. ADM.15/10/1/107 dated August 3, 1955, and entitled 'Tribal set-up', says: 'Isiolo District is composed of two distinct parts:
>
> 1. The Isiolo Leasehold Area, including for all practical purposes the Isiolo Township.
> 2. The Boran area – The Isiolo Leasehold Area is the western end of the district and ends at Chandler's fall on the Uaso Nyiro well above Merti. It is entirely on the south side of Uaso Nyiro. The Boran area is the eastern end of the district – i.e. east of Chandlers Falls and extends both north and south of the Uaso Nyiro.'
>
> The letter says that the residents of the Leasehold area are Somalis and Turkana. On the former it says, 'The permanent residents are almost entirely Isaac [Isaaq] and Herti Somalis'. On the latter it says, 'I have omitted to refer to the Turkana' [. . .] The letter even refers to Somalis from Wajir. It says: 'As you will realise most of the Somalis who originated from Wajir and now live in Isiolo are on semi-permanent passes'. [. . .] The existence of these groups is further defined in Somali commonage proclamation No. 21/36. According to the letter, the Borana were confined in the Boran area of the district. [. . .] The claim often repeated by Borana politicians and now the minister that Isiolo central and its environs belongs to the Borana community which includes the Leasehold area lacks historical validity.[. . .] Isiolo District belongs to all Kenyans and any Kenyan has the right to settle in the district. Ethnic cleansing of Somalis or other groups will not serve the interest of Kenyans [. . .]. Regarding the absurd claim about insecurity made by some Borana politicians, the lawmakers should know that the ethics of civilised society and Kenyan statutory laws call for the prosecution of a suspected criminal in court and not killing, robbing, demonisation, or expulsion of the criminal's tribe.

The MP Mokku, however, did not go into such details and continued:

> *Sisi viongozi, tumesema na tutaendelea kusema, wale watu wamekuja hapa hivi majuzi kwa malisho na sasa bunduki yao zimetatiza wakaaji wa hapa, waondoke na warudi kwao. Bunduki yao zimetatiza uchumi.* County Council *ya hapa ambaye mapato yake ni* revenue *ya shilligi milioni ishirini, na wanapatia watoto wa shule* bursary *ya milioni kumi, hivi sasa wameshindwa kulipa mishahara ya wafanyi kazi.*

101

We leaders have said and will continue to say that the people, who have come here for pasture recently, now that their guns have disrupted the life of the residents of this area, should leave and go back home. Their guns have disrupted the economy. The County Council of this area, whose income used to be a revenue of twenty million shillings, and gives the school children bursaries of ten million shillings, is now even unable to pay salaries to it workers.

Hayo yote imeletwa na wageni ambayo wamekuja hapa kwa malisho. Hili jambo, tutaendelea kusema. Na kama ni hili jambo ambaye itaendelea kula Wajumbe mojamoja, au itaendelea kuua mtu, iendelee hivyo hivyo. [. . .]. Sisi viongozi Waborana hatujasema ya kuwa jamii yeyote ambaye ni wakaaji na wazalendo, waondoke, kamwe hatutasema hivyo.[. . .]

All this has been brought by the strangers who have come for pasture. These matters, we shall continue to speak about them. Even if it is this matter which will continue to consume ('eat') one Member of Parliament after the other, or if it continues to kill people, let it continue like that. [. . .]

Sisi wakaaji wa Isiolo tumepiga kura kwa wengi kwa serikali ya KANU. Lakini siku hile tunalalamika kueleza shida letu ni kiboko. [. . .]. Hivi sasa ninawaomba wakaaji wa Isiolo, mtulie na mkae na salama. Sisi wenyewe tutatatua shida letu. Mtu yeyote asijaribu kulipisha kisasi.

We, the residents of Isiolo voted in majority for the KANU government. But the day we complained and expressed our problems it was the hippo whip. [. . .] That is why I now ask the residents of Isiolo to be patient and to stay peaceful. We shall get rid of our problem ourselves. Nobody should try to seek vengeance himself.

Ningependa kuwa kuwakumbusha tangu mwanzoni, marehemu PC Chelanga ameamuru wafugaaji hao waondoke. Hawajaondoka. PC Mberia pia ametoa amri wa wiki mbili waondoke, bado wako mpaka wa leo. Delegates mara mbili wameenda Embu na Nairobi kulalamika. Hakuna hatua yeyote ya serikali. Sasa sisi wenyewe wakaaji wa Isiolo tutatatua shida yetu pamoja.

I would like to remind you that from the start the late PC [Provincial Commissioner] Chelanga ordered the pastoralists to leave. They have not left yet. Also PC Mberia gave a two-week notice for them to leave, but they are still here until today. Delegates have gone twice to Embu [the Provincial Capital] and to Nairobi to complain. There was not action at all by the government. Now we ourselves, the residents of Isiolo, shall remove our problem together.

There was one comment in the same paper around this time by a journalist, Mutuma Mathiu, who had taken an anthropology course at University and was not satisfied with what he had learned there. He deconstructs in the light of the Isiolo events the functionalist view of ethnic cohesion:

'These lords of tribalism are Africa's curse'

[. . .]Were it not for the tragedy of a people wasting their time and resources in a stupid cause, one would be intensely tempted to sit back and enjoy the comedy of the all too frequent tribal skirmishes, for there is plenty of comedy just beneath the tragedy (or is it tragedy beneath the comedy?).

Now, ethnic identification – manifest, in most respects, as tribalism – is one of the most bogus concepts I have ever come across. I have heard many arguments over these many years, mostly from haggard intellectuals, that ethnicity can be a boon for nationalism. These theories – in the warm confines of Lecture Room 17 [. . .], the course being Anthro [sic] 101 – taste very theoretical when you place them in the real world of Kisangani, Srebrenica or Belfast.

For if there is anything I have proved beyond the faintest shadow of a doubt – and I dare anybody, the haggard intellectuals included, to challenge me on this – it is that ethnic identification has never brought any good to any society known to mankind.[. . .]

The problem in Isiolo is not scarcity of resources, real as that is. Neither is it that the Somali have invaded another district. The problem is that local leaders believe that whatever resources are available should be shared not on the basis of objective competition, but on the basis of tribe. [. . .]

When a leader argues that Kenyans are free to make a living in any part of the country, this is invariably not an expression of allegiance to the concept of a nation in which all are equal and free. It is another way of saying Kenyans have a right to own land and businesses anywhere in the country so long as they don't come to 'our' region.

Neither Mr Charfano Mokku nor those who oppose his views can conceive of Kenya as being 'ours' in a corporate, collective sense. They see it as a composite bastard, made up of segments, each owned by the Turkana, Marakwet, Kikuyu, Somali or whatever – with members owing priority allegiance to their particular segment. [. . .]

And how are we going to crawl out of this hole? Well, don't ask me difficult questions. All I can say is that tribalism is the ideology of the lazy and those without ability – those who seek free, easy things. It is the mother of all corruption, the dunderheads in high places and the border-line cretins driving stretch limousines, the mother of a wasteful regime of inequality and mediocrity. [. . .] (*Daily Nation*, 21 May 2000 p. 7)

In a letter to the editor, by a person with a Boran name in the USA, a decidedly pro-Boran perspective is taken, accusing the Somali of expansionism and wasteful resource management. The position is not new. Baxter, who did his field research in the 1950s, told Schlee about the very same accusations which the Boran levelled against the Somali in that period. There seems to be some truth in this position.

'Genesis of the Isiolo conflict'

To understand the current problems in Isiolo, one has to look back at the history of the region. If you start from the colonial era, it is easy to see a clear trend and establish who the real aggressors are.

First, there was the westward expansion by the Somali, which resulted in their encroachment on Borana grazing lands. Then there was the dream of the Greater Somali which brought about the devastating *shifta* war in the 1960s. Now we are back to the same old story – conflict over pasture.

Although this problem came to the limelight just recently, the Borana have been subjected to a reign of terror through constant banditry for more than 30 years in the 1970s and 1980s when Somali bandits raided Borana areas and stole livestock, they simply moved the animals to the neighbouring districts. These animals were rarely recovered.

Many Boranas were killed during those years, but the government did nothing to stop these continuous, deadly incursions by Somali invaders. The invaders have little regard for the traditional Borana grazing laws (*Seera dhedha*) and have destroyed most of the pastures. It seems the government has now realized that there is a crisis. What a mockery of justice after so many lives have been lost!

During droughts, the Borana used to have access to Gotu in Isiolo, Shaba and some areas close to the Meru border. Some of these areas are now reserved for wildlife.

The whole of northern Kenya has similar climatic conditions. The difference, in my opinion, is in traditional resources management strategies. The Borana take full responsibility for their grazing land. They devise a grazing strategy that ensures some tracts of pastureland are left idle until the next season. This restores the viability of the land.

If Somali nomads move to Borana areas and find some areas unoccupied, it does not mean that these have been abandoned. So when the Borana return to their original lands and find them occupied, what do they do? In most cases, the land is no longer inhabitable. It is this constant encroachment that ignites friction and bloody conflicts.

It is common to hear people say, well, northern Kenya is 'trust land'. Some of them will even call the area 'no man's land'. The use of such a term is misleading. The land is not free for all.

The boundaries set during colonial times, though not perfect, were meant to protect the communities within. However, after independence, all this changed; pastoralists no longer confine themselves to their own territories.

Why do the Borana live in fear of losing ownership rights over their own land? Three decades have gone by since the country got independence, and it is sad to see many innocent lives lost. Is the threat to Borana lands real or imagined? The evidence is clear, but I wonder whether the current law will protect them.

I do not suggest that the Boranas or any Kenyan citizens take the law into their hands. There is no justification for any senseless killings. But something must be done.

Dido G. Kotile,
Illinois, USA.
(*Daily Nation*, 22 May 2000 p. 7)

The problem addressed by the letter writer is that known in social theory as 'the tragedy of the commons' (Hardin 1968). Here this theorem is applicable in its quite literal sense as we are dealing with common grazing lands. In the case of industrial societies it is mostly applied metaphorically to other types of freely available resources like air and water which are overused or polluted because there are no adequate incentives for their preservation. Among themselves the Boran appear to have mastered the problem of overuse of shared resources by giving themselves rules (for a study of resource management among the related Tana Orma see Ensminger 1992), but the Somali apparently do not adhere to these rules, thus enjoying the extra benefits of those who do not participate in common efforts and do not follow the rules, something which is known to economists as the 'free rider effect'. This argument, however, restricts

itself to one field of resource competition, namely pastoralism, while the

Isiolo conflict appears to have also purely urban layers and also seems to derive some of its momentum from the level of national politics.

We have seen that the rural, pastoral facets of the conflict in Moyale involved people from Wajir District and areas of that District. Soon the mutual raiding going on there was also to affect Kenyan/Ethiopian relations. On 14 June 2000, the *Daily Nation* reports on p. 9:

'Wajir residents protest militia attack'

> Members of the Ajuran clan yesterday took to Wajir town streets to protest a recent attack by militiamen from Ethiopia. Waving placards and chanting slogans, the Ajuran accused the Garre clan of masterminding the raid in which two people were killed and more than 10,000 head of cattle stolen last week in Wajir North. Led by councillors Abdulrahman Tebo and Ibrahim Waso, the residents demanded the immediate return of the stolen livestock.
>
> Wajir DC, Mr Fred Mutsami, said the Government has lodged a formal complaint with the Ethiopian authorities over the attack.
>
> The DC said the Ethiopian authorities had requested for 10 days in which the stolen livestock will be returned.

Under the headline 'Mission to Ethiopia unsuccessful' the *Daily Nation* reported on 24 June 2000 (p.24) that Kenyan leaders had only been offered the return of 160 animals which Ethiopian forces had confiscated from the raiders.

> The delegation, led by the Wajir DC, Mr Fred Mutsami, talked with Ethiopian authorities at Gadaduma on the border on the return of more than 6,000 animals stolen from Wajir North.
>
> [. . .] He added that the Kenyan delegation, which included the Wajir North MP, Dr Abdullahi Ibrahim Ali, refused to accept the animals, and insisted that the entire herd be returned.

In fact, the numbers given by the Ajuran might have been exaggerated. What appears actually to have happened is that Garre from Ethiopia had raided Kenyan Ajuran. Some of the raiders wore uniforms. They might have been soldiers on an unauthorized mission, an off-duty income-generating activity, so to say.

The Ajuran, in order to mobilize official Kenyan support, however, laid the stress on the Ethiopian involvement, thus trying to transfer a *trans*national affair between Kenyan and Ethiopian herdsman to the *inter*national level. In the *Daily Nation* of 24 June 2000 (p. 7), we read:

'Ethiopian soldiers harassing us'

> We, the elders of the Ajuran community, strongly denounce recent atrocities committed by regular Ethiopian army soldiers on defenceless civilians of our community in Kenya.
>
> Some herdsmen were looking after their livestock in Wajir North Constituency when the soldiers attacked killing two, wounding two and driving away 5,000 cattle and 900 camels.

Since Kenya's independence, the ties and relations between the Kenyan and Ethiopian governments have been cordial. But recent developments [. . .] have made this historic relationship unworkable. The government needs to make relevant foreign policy adjustments to [. . .] counter Ethiopia's aggression.

We earnestly appeal for President Moi's personal intervention on this matter so the Ajuran community can repossess their livestock. Unless this happens, our government will lose [. . .] legitimacy [. . .]. It is the government's primary responsibility to protect the lives of its citizens and their property.

Also the Provincial Commissioner of the North Eastern Province held Ethiopia responsible for this incident:

Journalists who accompanied the PC during the tour were shown more than 20 houses that were burnt down during the four-hour raid. The buildings are located a few metres from the local police station. The journalists were also shown the body of one of the raiders who was killed in the Melee. (*East African Standard*, 28 June 2000 p. 9)

The Ethiopian embassy in Kenya denied sharply that Ethiopia had anything to do with the affair:

Ethiopia fully respects Kenya's sovereignty and strongly values the shared bond of friendship. The claims, that Ethiopian soldiers have been making incursions into Kenya and killing innocent civilians, are outrageous and not in keeping with the friendly relations between the two countries. (*Daily Nation*, 1 July 2000 p. 7)

As if education in northern Kenya was not ridden with enough problems in normal times, the insecurity led to the closing of schools. The *Daily Nation*, 7 July 2000 (p. 19), reports:

District Education Officer Abdi Badel yesterday confirmed that two primary schools had been closed because teachers and pupils had fled for fear of attack. [. . .]

There has been tension in Wajir North for the last three weeks, following fresh animosity between the Garre and Ajuran clans. The feud has so far claimed the lives of more than 32 people [. . .]

Some of the pressure which local leaders had been putting on the Government to take responsibility for the security situation in the area was turned around, as Government officials blamed the local MPs for provoking the clashes:

'Government probes MPs over clan feud'

[. . .] North Eastern PC Maurice Makhanu yesterday said Wajir North MP Abdulahi Ali and his Mandera West counterpart, Said Mohammed Amin, have been accused of inciting Garre and Ajuran clans to war that has claimed over 30 lives in the last one month.

Saying that the MPs were not above the law, the PC instructed police to arrest them if they make any inflammatory statements likely to fan the inter-clan feuds [. . .]

But when contacted, the MPs accused the PC of being responsible for the clan feuds and called for his immediate transfer from the province.

As evidence for his inefficiency his stand on the OLF presence in the area, which has been discussed above, was recalled. The MP Amin said Makhanu had been denying the presence of the Oromo Liberation Front 'rebels', even after reports by the locals about them. As further evidence for his lack of control it was adduced that 'since he was posted there, clan feuds have escalated in Garissa, Wajir and Mandera Districts'.

> Amin said last week's killings of 20 people in El-Danaba trading centre was a result of laxity by the administration because the victims had earlier appealed for security.
> Makhanu announced that four senior chiefs have been suspended indefinitely with immediate effect for compliance in the clashes. He said the chiefs, three of them from Mandera District and one from Wajir, were abetting proliferation of illegal firearms to support their clansmen and were inciting them to war.
> He identified the suspended chiefs as Abdisalan Ibrahim (Bute), Isaack Alow (Elwak), Isaac Garrow (Borehole 11 location) and Mohammed Soba (Kutulow). The PC said the clan clashes were politically instigated and were fanned by politicians. (*East African Standard*, 9 July 2000 p. 28)

At least as far as Abdisalam Ibrahim, the Ajuran chief of Bute location and the brother of Dr Abdullahi Ibrahim, the MP of Wajir North (KANU), was concerned, this suspension has remained a mere threat. He remained in office until his retirement in 2002 after 37 years in government service, first as an Administration Police Officer, and from 1978 as a chief. He is still (in 2011) active as the chairman of the Wajir North Peace Committee. The article continues its report about the speech by the PC:

> He said a leader's meeting will soon be convened to discuss the issue. (*East African Standard*, 9 July 2000 p. 28)

This leader's meeting, to discuss reconciliation of Ajuran and Garre, was first planned to take place at Wajir. But the Garre delegates refused to go there out of fear for their lives. Then it was to take place at Mandera, where the Ajuran refused to go. The next proposed location was Garissa, the Provincial capital, where neither the Garre nor the Ajuran felt secure. Finally Moyale, although it is situated in another province, Eastern Province, was agreed on as neutral ground. The meeting came as a surprise to Boran residents of the area who were infuriated by it. They feared that the problems of the neighbouring province were exported to them.

The meeting took place on 27 June 2000. The plan was that after a closed session by the Provincial Security Committee, which consists of Government officers only, separate talks would be held with the Ajuran, then with the Garre, and then a joint session with representatives from both groups would be held. However, an éclat occurred relating to the Ajuran part of the programme. The Ajuran delegates walked out after 107 declaring that they had no time for meetings announced at short notice, because they had to bury their dead whom the Government had failed to protect and to retrieve their looted livestock which the Government had been unable to do. There was no time for peace with the Garre now. Confronted with accusations that they received help by the OLF which is alleged to have returned to the area, the delegates responded provocatively in the affirmative: yes, even the OLF units from the Sudan would rush to their help. The Blue Nile area of the Sudan, where the OLF once had bases, is about 1,000 kilometres away from northern Kenya.

On the national level, the insecurity in the north was used by the opposition to criticize the Government. One can well imagine that regional administrators like PCs and DCs were in an uncomfortable situation between angry local elders and impatient superiors in Nairobi. The head of the largest opposition party made a press statement on 17 July 2000 (*East African Standard*, p. 8):

'Government urged to negotiate on security'

Leader of Official Opposition, Mwai Kibaki,[65] has asked the Government to immediately enter into talks with Ethiopia so as to end massacres and cattle rustling in North Eastern and Eastern provinces.
Kibaki wondered why the Government was quiet over the killings and loss of property that have rocked the area in the last three months. He said there was no need for President Moi to mediate between Eritrea and Ethiopia while his territory is suffering similar problems. [...]

This pressure on the Government was increased by press comments like the following:

'Clan wars: Who'll stop the carnage?'

The senseless killing of 35 people in inter-clan fighting in Wajir these past two weeks is only the latest outrage in a continuing orgy of violence that has engulfed northern Kenya for too long.
In the past six months alone, 120 people have been killed in Isiolo, Garissa and Wajir in 15 violent inter-clan confrontations. These are only figures the Government has confirmed. They have been vehemently disputed by local leaders as a gross understatement. [...]
There is never any shortage of Government reassurances about security being beefed up and the attackers being pursued, even as the mayhem and slaughter continue unabated. Often, the army is deployed in these trouble-spots [...]. Success in this strategy is yet to be fully evaluated.
Even as the Government is accused of wilfully neglecting the entire region in the face of natural and man-made catastrophes as well as the rising insecurity – a charge that cannot be easily brushed aside – it is necessary to note that the areas in question border highly volatile, war ravaged countries.
This factor alone has contributed at a great deal to the culture of violence. The Government has been forced to arm local residents as police reservists, if only to provide a measure of security for the residents.

---

[65] Mwai Kibaki was elected President of Kenya in December 2002. During his own tenure, however, he has hardly been more concerned with northern Kenya than his predecessor. He did not, for instance, visit the victims of the Turbi massacre (see below).

The militarisation of these homeguards has made it easy for them to sustain the incessant fighting. It is high time an alternative security system was considered and implemented.

What remains indisputable is the patent reality that the Government has lost grip of the security situation in northern, eastern, and North Eastern Kenya. It is unimaginable that a Government can condone this kind of anarchy. These regions are fast acquiring the image of ungovernable bastions run by untouchable clan warlords. It is an image the Government, any government, cannot allow to solidify if it believes that it is in charge. (*Daily Nation*, 22 July 2000 p. 6)

There were rumours in the area that the Garre had hired militiamen, Garre from Somalia who had joined the RRA (Rahanweyn Resistance Army). These rumours are also reflected by the following press report. It is also remarkable that tribal territories are described as if they were legal entities or naturally given:

'Clan clashes toll rises to 35'

Police yesterday confirmed that 35 people were shot dead when two rival clans clashed in Wajir District. Those killed included 30 members of the Garre clan and five elders from the Ajuran clan.

However, it was clarified that violence broke out when about 150 gunmen from the Garre clan invaded the territory of the Ajuran.

The raiders from Ethiopia and Somalia, backed by their clansmen from Mandera District and armed with AK-47 assault rifles and sub-machine guns, crossed the Kenya-Ethiopia border and launched the attack on Tuesday.

[. . .] Mr Omar [the county councillor for Gurar] said the raids were aimed at displacing the Ajuran from the area to 'facilitate the easy smuggling of illegal firearms into Kenya'. 'Members of my clan, the Ajuran, are opposed to arms smuggling and that is why they are being massacred', he said. (*Daily Nation*, 22 July 2000 p. 24)

On 4 to 6 August 2000, eleven Members of Parliament from all over North Eastern Province met at Wajir to hold a public *baraza*[66]. Under the headline 'Support Somalia peace process, MPs tell the Government',[67] the *East African Standard* reported on 7 August 2000 (p. 5) that the MPs urged the Kenya Government to support the on-going peace initiatives in Somalia in order to curb insecurity in the region. They noted that break down in law and order in Somalia was the main cause of insecurity in the region.

[. . .] They said the government should have taken a leading role in the Somali peace talks that are being held in Djibouti. 'North Eastern [Province] will not have peace as long as there are open air firearm markets in neighbouring Somalia', they said. [. . .] They said Kenya hosted thousands of Somali refugees and should therefore pay attention to goings on [with] its eastern neighbour.

Foreign Affairs Assistant Minister, Mohammed Abdi Affey, promised to take up the matter with his boss, Dr Bonaya Godana. Meanwhile, the

[66] Sw. public meeting.

[67] For a summary of the Somali Peace Process at Arta, Djibouti, and subsequently at Eldoret and later Mbagathi (Nairobi), Kenya see Schlee (2006, 2008a).

leaders amicably resolved the long standing dispute between the Garre and Ajuran clans of Wajir North.

A meeting held at Wajir Girls Secondary School under the chairmanship of Cabinet Minister Hussein Maalim and Assistant Minister, Yussuf Haji, saw leaders from the two warring clans resolve to end the clashes immediately.

Elders from the two communities, led by Abdullahi Ali of Wajir North and Said Mohammed Amin of Mandera West, unanimously agreed to abide by all the 15-point blue-print drawn by leaders from Garissa, Ijara and Mandera.

Among the resolutions were the inclusion of disputed areas of Iresteno [Irres T'eeno, see above] and Sagemada [Eel Danaba?] into Wajir North and the immediate resettlement of all the displaced families from Bute, Danaba and Gurar.

This last point, the inclusion of the disputed areas into Wajir District only confirmed the official boundaries. What had happened there, as we have mentioned above, is that the Garre had claimed the posts of Chief and Assistant Chief there. If they had achieved this aim, both in local perception and in administrative practice the area might then have been regarded as a part of Mandera District. A parallel example is Turbi location, which, according to the map, belongs to Moyale District, but which is treated as a part of Marsabit District because all office holders there are Gabra Malbe. The Garre were expelled from all of northern Wajir District. In the towns of Bute, where the Garre had eighteen permanent houses, and in Gurar, where they had six, the Garre abandoned their properties. Their houses were dismantled by the Ajuran, and the corrugated iron sheets and concrete building blocks were taken for other constructions.

The 'amicable resolution' of the 'dispute' between the Ajuran and Garre was agreed on at 2 p.m. The very same evening bombs exploded at Bute and Gurar.

The Minister, Hon. Hussein Maalim, MP for Garissa, at this meeting, gave emphasis to the resolutions passed (in Swahili, in the words of Adan Jillo, a listener in the public *baraza*, who gave Shongolo a telephone interview the same evening):

> [. . .] *Lazima tuwache vita kwa sababu kama sisi waislamu tunavunja amri ya Mwenyezi Mungu. Imekuwa tabia ya sisi wakaaji wa North Eastern kuingia sehemu ya Waborana kutafuta malisho wakati wa kiangazi. Waborana wenyewe hawaja ingia sehemu yetu wakati wowote, hata wakati wa hali hatari ya kiangazi. Waborana wakikaribisha wafugaaji wetu, ninyi wenyewe mnapigana juu ya ardi ya waborana. Wafugaaji wanatoka sehemu hii wakiwa matajiri ya mali na kisha kurudi hapa katika hali ya umaskini.*

It is necessary that we stop the war because, all of us being Muslims, we are breaking the orders of God. We, the residents of North Eastern Province, usually went into the part of the Boran to look for pastures in the time of drought. As to the Boran, they never went into our part at any time, even in an emergency situation in a severe drought. And when the Boran welcome our pastoralists, then you yourselves start to fight on Boran soil. The pastoralists leave our area rich in livestock and later come back poor.

[. . .] Ni sisi wenyewe tumewachokoza Waborana kwa miaka nyingi. Lazima sasa tuwachukue Waborana kama ndugu, kwani ni sisi ambaye tuna haaja na wao mara kwa mara. Lazima pia tuhudumishe ujirani mwema na wakaaji wa serikali ya Ethiopia. Vita tuliopigana hapa Kenya *zimeenea kule na kusababisha maafa nyingi [. . .] Tujaribu kusameana na kurudisha usalama kwa jina la Mwenyezi Mungu.*

It is us ourselves who annoyed the Boran for many years. We now have to accept the Boran as brothers, because it is us who are in need of them from time to time. We also have to maintain good neighbourliness with the local populations who belong to the Ethiopian government. The war which we have been fighting here in Kenya spilled over and caused many deaths. [. . .] Let us try to forgive each other and to restore peace in the name of the Almighty God.

As the battles between the Ethiopian Garre and the Kenyan Ajuran intensified from early June 2000, those between the Ethiopian Boran and Gabra Miigo on the one hand and the Garre intensified as well. On 26 June 2000, a meeting was held by the Ethiopian government officials in Moiale. Representatives of the Gabra, Boran and the Garre narrated the causes of the conflict. A Gabra elder depicted the interethnic relations with the Garre:

[. . .] Garrin waan chufaan nu d'ibde. Warri aada womuutu gul hin bulu. Aada Isilaana, ta Borana, yokaan ta sirkaala tokkolle gul hin bulu. Laf teen yo waliin qubannu, warri waan jed'u, lafti ta Waaqa, jed'a. Ta warra yo waliin qabannu, teenum ta Garrii, jed'a.

The Garre have disturbed us in every way. They follow no custom. Be it the Islamic custom, the one of the Borana or the one of the Government, they do not follow any of them. If we settle together in our land, they say that the land belongs to God. If our people settle among them, they say "it is ours", of the Garre.

Marra bisaan nu d'owwe, ufumaati soorata. D'io tan, birri d'ibba nurra fuuti, yo horiin bisaan seer isii keesatti obaafnu. Dir kees, bulti dad'amne. Nu dur hin deemtu, nu gul hin deemtu. Finni isii finnum Somalii, ka babbadaa [. . .]

They refused pasture and water to us, they eat alone. Recently they have taken a hundred Birr from us, if we water our animals within their boundaries. In the town, we can no longer sleep. They do not walk ahead of us, they do not follow us. Their way of life is that of the Somali, disorderly.

A Boran elder illustrated his speech with a proverbial conclusion:

[. . .] Duri wa sadi d'aadatte. 'Qeeramsi d'aadate waan jed'e, yo an wa gaadu danda'ani sokorsa tiy hin d'aga'ani. Wanni an qabu, dandeete narra hin baafattu. Yo isiin narra baafatte, nattu ilme qeeramsa hin taini.'

Long ago, three made a declaration. The leopard declared: 'If I ambush something, no one can hear my approach. What I get hold of, cannot get away from me. If it gets away from me, then I am not the son of a leopard.'

Nenchi waan d'aadate, 'ejjata faan tiyya hín d'aga'ani. Ammo wanni an gaadu, dandeete na dura hin baatu. Yo baate, nattu ilme nencha hin taini.'

The lion declared: 'my footsteps can be heard. But what I ambush, cannot escape me. If it escapes, then I am not the son of a lion.'

Garriillen waan jette, 'lafa nu qubanne itti waggine, nuu qara hin gogodaanna, ammo fula nu itti waggine, yo isiin naga-aan teete, nuttu ilm Garrii hin taini'. Akkanaafu, Garri bultiin isii rafuji kees jirti. Laf tan, waraan itti buufte, duduubatti rafuji taati, bulti harka sirkaala irra fed'ati.

And the Garre said: 'the land in which we settled and lived for a year, we normally migrate, but where we stay for a whole year, if peace prevails, we are not sons of Garre.' Therefore in the life of the Garre there is something of the way of refugees. They bring wars to this land and later become refugees, they seek livelihood in the hands of the Government.

Nu ammo warra qubate hori tiffatu, oobru qotatu, lafum tanaati d'alanne, akka warra gari itti hin d'umne, naga feena, marra bisaan hori keenna feena [...]

But we are people who have settled and look after our livestock, till our fields, have been born in this land, we have not come from elsewhere like these people, we want peace, we want pasture and water for our livestock.

In response to the allegations made against the Garre community a Garre elder, responded:

[...] Wanni nu himatan kun chufa d'ara. Yo nu Gabra Boran irra wa yakkine, waari nu gaafacu hin danda ka sirkaalan nu barbaanne. Nu ammo nagumaan warra chufa waliin taau feena [...]

What we are accused of is all lies. If we spoil anything which belongs to the Gabra and the Boran, they themselves can ask us about it without taking recourse to the government. But we want to stay in peace with all of them.

It is obvious that the conflicting parties were competing for the sympathy of the Government and its support. This round seems to have been won by the Gabra and Boran. In response to all, a senior government official concluded (translation of the Amharic speech):

[...] I have heard all your sentiments. But one thing is certain: that the Garre community are problematic people. Today some Boran still are supporters of the OLF but even though, whenever problems arise, we summon their leader, the Abba Gada, and hold him responsible. Like the Boran, the Gabra too have a Talia [title of an office holder]. He too is answerable for the misdeeds of his Gabra people. But with you, the Garre, you have no leadership. You have no head. Every individual is his own leader. The young do not obey the elders. You have no defined culture which binds you together or binds you either to Boran customs of this land or to the government rule of law. We have repeatedly advised you to desist from warlike habits. It is now up to the government to take action against those found breaching peace. This country is not like some other countries where one causes chaos and gets away with it, the way some of you did in Somalia. You should live at peace and never deprive others of their rights. [...]

The above meeting did not stop the war. Following bitter clashes between the Ethiopian Boran/Gabra and Garre in the last weeks, where many people were killed and livestock looted, the president of Oromia State (Region 4), and OPDO chairman, and the president of Region 5, the Somali region, and at the same time chairman of the oppositional Somali League, held a meeting in Moiale on 6 July 2000.[68]

In attendance were officials from the Prime Minister's office, in charge of internal security, the Provincial Security Team and the party officials for both Regions 4 and 5, the outgoing Abba Gada, Boru Mad'a, and the incoming Abba Gada, Jaldes Liiban, the Gabra Talia, Hassan Qalla, representatives of the Boran, the Gabra and the Garre communities. In summarizing the allegations, the Abba Gada, Boru Mad'a said:

> [. . .] *Akkum duri yo boni d'eeratu, marra bisaan barbaannu, Borani arra fula afuriiti gargari bae. Tokko Burjiin bae, tokko Gujiin bae, tokko Konsoon bae, kaan armaati Garriin bae. Ka kara kaan sadeeni, Gabra Borani naga qaba ammo ka kara Garrii kun, waraan qaba.*

[. . .] As in the past, when the drought becomes long, we look for pasture and water, and the Boran have now split in four parts. One went to the Burji, one to the Guji, one to the Konso, and the other came here to the Garre. On the three other roads the Gabra and Boran have peace, but the ones on this Garre road have found war.

> *Garriiti Gabra Boran marra bisaan irraati had'a. Lafaan teen jetti. [. . .] Jeshi sirkaala nutti fud'atte, tanaan, nam nurra hobaafti. Nami isii ka jeshi ka waraan keesatti due, kurnya hin ga. Milikeedi hid'ata isii kana [. . .]*

The Garre attacked the Gabra and Boran because of pasture and water. They say 'the land is ours'. [. . .] They brought the Government army against us and thus finished off our people. Their own people in the army, who died in this war, are about ten. The evidence for this is these uniforms. [. . .]

> *Raaya wal haatu beenne, sirkaal raaya ufi had'u, nu tan akka himnuun hin beennu. Kuno d'uftani dandeetan fala [. . .]*

We know about civilians fighting each other, but we do not know what to say about the government fighting their own civilians. Now that you have come, solve this problem, if you can.

A Garre elder gave his view pointing an accusing finger at the Kenyan pastoralists:

> [. . .] *Nu Gabra Boran Topiaatin wal hin d'abne. Horiin arra laf tan qabate ka Gabra Boran* Kenya [failed to mention the livestock of the Garre which were driven there earlier] *Marra bisaan nuhu Topiaatu walti hin gá'ani. Gabra Boran Kenya ammalle waiit gul yaa. Tanaafi marra bisaan d'owwanne. [. . .]*

[. . .] We did not quarrel with the Gabra and Boran of Ethiopia. The livestock which fills this land is that of the Gabra and Boran of Kenya.[69] The pasture and water is not enough even for us Ethiopians. And something is following the Gabra and Boran of Kenya [an allusion to the OLF]. That is why we denied them pasture and water. [. . .]

---

[68] Interview by Abdullahi Shongolo with Mr Nurow, a delegate in the meeting.

[69] Also the livestock of Kenyan Garre had been moved there before.

Warri duuba nu had'e. Waraani arra kuni, marra bisaan challaaniti, falama lafaati duuban jira. Gabra Borani, laf teen qubate, duduubatti falamu fed'a. Tan duraanu beekanne, dandeenne itti hin lakkimnu. Ammo nageenni namum chufaa irra jira, nu gar keenan naga la fud'anne [. . .]

Then they attacked us. The war of today is not only about pasture and water, there is a claim to the land behind it. The Gabra and Boran who settle in our land, will want to claim it later. We have come to know about it, and therefore we cannot let them in. But peace is good for all, as for us, we have accepted peace.

The officials from Addis Ababa, who had so far been mainly listening to the discussions, then concluded the meeting with a rather general appeal to all to maintain peace. They appear to have been briefed that the matter was all about an OLF invasion backed by Kenyan Boran and Ajuran, and were surprised to find a multi-layered conflict between the various communities within Ethiopia as well. They invited the elders to further meetings to solve these problems.

On 10 August 2000, it was reported by the Kenya Broadcasting Corporation that 65 Garre had been killed by Boran two days earlier in Ethiopia, at Arero, between Negelle and Yavello. Boran elders and Ethiopian government officers give the number of dead as 48, and that of the wounded as 31. Not all of the victims were Garre. The area was one in which informal gold mining took place. Here Garre had killed four Guji Oromo days before. A combined Boran/Guji force then ordered the Gabra, Boran, Burji and Guji to vacate the area. The Garre, who were not informed of this, and everyone who disregarded the order, then came under fire.

In 2001 the Ethiopian authorities discussed the possibility of a referendum as a means to stop the Boran/Garre conflict. In the area at the south-eastern fringe of the Oromia state, between Moiale and Negelle, people would vote whether they want this area to continue to belong to Oromia or to be transferred to Region 5, the Somali region. The Boran, however, wanted the war to end first and then everyone to be allowed to return to their prior residence, because many Boran have been expelled by the Garre and, according to them, a referendum would only legalize the results of the recent clashes. They also want their Abba Gada to be involved in the process. At the time (early 2001) this was not possible because he was occupied with rituals leading to the handing-over of his office to a new holder in early 2001[70]. In August 2000 he was in the Arero region performing the *oda* ceremony which involves sacrificial slaughters and prayers for the *hayyus* and other officials of the age-set which had newly moved into the *gada* grade. Also his own successor has been chosen and was moving around with him. Such ceremonies were held in different places all over Boranland. He had a heavy escort of Ethiopian government forces.

[70] A referendum eventually took place in 1994 – see chapter 2.

2

The Post-Moi Period
2002–2007
GÜNTHER SCHLEE & ABDULLAHI A. SHONGOLO

The later years of the Moi period brought an acceleration of the ethnicization of politics in Kenya. Politicians came to treat their constituencies as
ethnic territories and to behave like ethnic leaders. Former pretences at
modern statehood and universal citizenship were dropped to an increasing
degree, at least in practice and in speeches given and discussions held in
Swahili or local languages rather than in English. The change of government in December 2002 did not bring the change people hoped for. Ethnic
clientelism remained the key tool of politics. It is therefore justified to deal
with the late 1990s and the year 2000 with some detail, because it is this
period in which ethnic exclusionism became treated more and more as
normal and legitimate.

The proponents of ethnically pure sub-districts and districts and
special grazing rights for the titular tribes spoke of time-honoured rights
and traditional territories. As has been described in *Identities on the Move*
(Schlee 1989a) and *Islam and Ethnicity* (Schlee with Shongolo 2012)
however, there were no tribal territories in pre-colonial times. It was only
the British who created ethnic districts and grazing areas as a miniaturized
version of the European nation-state. It is these colonial boundaries which
now are evoked for giving legitimacy to territorial claims. In pre-colonial
times it was the elders of different groups within the Boran-centred *Worr
Libin* Alliance who discussed matters of grazing and access to water points
and the resulting pattern was not one of stable, neatly divided territories
but of spatio-temporal arrangements with shared resources in one season
and separation in another. With people not forming parts of this alliance
the occupation of areas was a result of mutual raiding pressure. The spaces
of movements expanded and numbers of options arose if a given group
could keep their hostile neighbours under pressure, and they shrunk when
these neighbours put them under pressure.

We have described some of the rhetoric giving legitimacy to ethnic
territoriality in the preceding chapter which has taken us up to 2000 and
was written in brief form after the events to which it refers. We then still
found the Gabra in an alliance with the Boran and the Boran establishment, most prominently the Abba Gada, in an alliance with the Ethiopian
Government. The pattern since then has changed drastically. Now there
is a deep rift between Gabra and Boran, both in Ethiopia and Kenya, and 115

it is the Gabra who make great efforts to depict themselves as loyal to the Ethiopian Government. The new Abba Gada of the Boran has taken a much more critical stance towards the government than his predecessor. We will, however, not be able to deal with the period 2000 to 2007 with the same level of detail. It is also no longer necessary to show how the principle of ethnic territoriality was rhetorically established on the Kenyan side, because, sadly, that had happened earlier and in this period everyone seemed to adhere to this disruptive premise. In Ethiopia, anyhow, ethnic federalism is part of the constitutional order since the early 1990s.

The Gabra/Boran conflict on the Kenyan side reached a culmination point on 12 July 2005, when 76 Gabra, many of them school children, were massacred by a combined force of Ethiopian and Kenyan Boran at Turbi (contemporaneous notes by Abdullahi Shongolo). This event can also be analysed in the context of territorialized ethnicity. One of its causes was an unclear boundary inherited from colonial times and unclear boundaries are no longer tolerated. Turbi currently falls under Marsabit District but the people of the newly re-created Moyale District and the Boran in general believe that Turbi was part and parcel of the colonial Moyale District, which later became a part of Marsabit District. While it was indeed part of Marsabit District, the boundary was one of internal divisions. Moyale community first raised the issue during the Waitate boundary Commission meeting in Marsabit in 1984. They claimed that Turbi should be part of the Moyale constituency and not of North Horr, an area predominantly occupied by the Gabra (notes by Abdullahi Shongolo, present at the time). This sentiment continued through the years with varying degree of intensity and rose to new heights when Moyale became an independent district again in 1996 and Turbi remained as part of Marsabit District. The Gabra claim that Turbi belongs to them but Boran believe that Moyale District boundary is at *Lag Warabesa* ('Hyena River', approximately 30 kilometres beyond Turbi to the south). They justify their claim by the fact that Mohamed Galgalo, the Moyale MP at the time, constructed the Turbi pan, which is currently the main source of water for the Turbi people.

The Turbi boundary has remained a contentious political issue and a very powerful campaign tool ever since for politicians from both North Horr and Moyale constituencies. Moyale politicians, including the late MP Dr Guracha Galgallo often pledged to bring Turbi back to Moyale while his North Horr counterpart, the late Dr Bonaya Godana, promised to retain the status quo. He was suspected of encouraging Gabra to settle there with the aim of making Turbi an exclusively Gabra area. From a Gabra perspective, Boran claims to Turbi were part of a plan to create a continuous Boran belt from Ethiopia through Moyale and Marsabit all the way south to Isiolo.

The boundary problem extends from Turbi to Hurri hills where the Gabra often express very strong sentiments against the Boran settlement in the area. They claim that Hurri hills is their wet season grazing area but is being taken over by the Boran and the Konso, whom they consider to be illegal immigrants from Ethiopia. This land ownership problem runs along the Turbi-Hurri belt and fuels endless violent conflicts.

The escalation started in August 2002 when one Gabra was killed at Turbi, presumably by Boran.[1] The Gabra took revenge by raiding nearby Boran herds and taking away 728 sheep and goats. Raids and isolated killings were characteristic for the next two years. The escalation process was interrupted by peace meetings, facilitated by a number of NGOs.[2] The results of these meetings, however, were either inconclusive or the compensation payments and restitutions of stock taken as loot on which these meetings decided were never implemented.

The conflict became less and less localized. The boundary area around Forole and the Hurri Hills were involved, and in January 2005 the conflict lost its hitherto predominantly rural and pastoral character when Gabra houses were burned and a Gabra man killed in Marsabit, the District capital. The Gabra then killed six Boran in their sleep near Forole.[3] That seems to have been the trigger for the real massacre, in which, as we have mentioned, 76 Gabra were killed at Turbi.

What set in now was an ethnic cleansing. A Boran resident of Maikona, the son of the woman who runs the local food kiosk, was intercepted on his way home from the mosque and had his throat cut. Like this, in many places people turned against their long-term neighbours and friends and killed them or, worse, tortured them to death. Children were taught brutality by witnessing such scenes or by being given tied-up victims as targets for stone-throwing.

Still, only the Gabra were presented as the victims by the louder political voices echoed by the media and thus sympathies were on their side. Politically, it was they who won. Immediately after the Turbi massacre the status of Turbi was lifted from that of a location to that of a division – a division of Marsabit District, needless to say, not of Moyale District.

Also from interviews with local people, it was clear that Marsabit District politics had an influence over the latest conflicts between the Gabra and the Boran communities. The fact that political leaders have access to Constituency Development Funds (CDF), Constituency Roads Funds, Constituency Bursary Funds, LATF (Local Authority Trust Fund) and LASDAP (Local Authority Service Delivery Assistance Programme), has led to an increased interest in controlling political processes in the district. The competitive nature of politics has created the need to secure predictable voting blocs, resulting into formation of alliances between ethnic communities. For instance, an alleged alliance between the Rendille, the Gabra and the Burji (REGABU) was formed as one voting block. The power of this voting bloc has been tried with success during Marsabit Teachers' SACCO[4] election where candidates supported by

[1] From Abdullahi Shongolo's notes taken at a meeting he attended in Moyale District.

[2] COPEP (Community Peace Programme), CIFA (Community Initiatives Facilitation Assistance), PISP (Pastoralist Integrated Support Programme), PARIMA (Pastoralist Risk Management).

[3] From Shongolo's field notes of an interview with elders in Marsabit.

[4] SACCO stands for Savings And Credit Cooperative Organization.

REGABU won. A Rendille/Gabra/Burji alliance leads, of course, to the exclusion of Boran.

In the recent past, the late MPs Abdi Sasura of Saku and Titus Ngoyoni of Laisamis constituencies have made attempts to disband the alliance. The latest conflicts can be seen from this perspective. The fear of an 'ethnic coalition' and block voting in the general election of 2007 was a political factor playing into the latest conflicts. The reaction against an ethnic coalition, however, tends to be nothing better than an alternative ethnic coalition.[5]

Up to his death in a plane crash in 2006 Hon. Bonaya Godana was a clearly dominating figure and that isolated him somewhat among his colleagues, the other MPs from Marsabit. Apart from trying to expand the areas of Gabra occupancy, he was suspected of favouring his Gabra tribesmen in education and employment opportunities. He was particularly well placed to do so during his tenure as Cabinet Minister (1992–2002), when he was first holding the key Ministry of Foreign Affairs, then the Ministry of Agriculture.

The political struggle about district boundaries and divisions within a district is linked to competition about pastoral resources and a struggle about political power. The Gabra area had suffered years of drought. While they had a strong leader, Dr Godana, famine relief and even water was brought to their settlements by lorry, so that the survival of people was guaranteed. What the Gabra, however, had to struggle for was the survival of their herds. They pushed towards all sides for pasture for their suffering herds, looting their neighbours to replenish their stock. They were fighting the Rendille, Turkana and the Dasanech at the same time as the Boran. While water and grazing is the herder's stake in the conflict, power on the national level is at stake for the politicians. This is vividly illustrated by the blame the (Boran) Members of Parliament for Moyale, Saku and Isiolo on the one hand and Dr Godana, the North Horr MP (Gabra), mutually put on each other for having instigated the conflict in the aftermath of the Turbi massacre.

It has often been claimed that the Gabra/Boran conflict could have been defused by the implementation of 'traditional' law. In fact the 'traditional' law in question is that resulting from recent attempts to standardize compensation payments following the Somali model, as laid down in the Madogashe[6] Declaration. This declaration recommends that 'once livestock is stolen, twice this number should be paid back by the offending community. In the same spirit, 100 camels are paid as compensation for every man killed and half the figure for a woman.' (*The Standard* Monday 18 July 2005 p. 23). 'This declaration [. . .] had helped to resolve several disputes in Marsabit. But its significance was watered down by President Kibaki during his visit to Mandera in January

[5] The two MPs formed BORE (Boran/Rendille) alliance against the Gabra.

[6] Madogashe is a locality between Isiolo and Wajir. A declaration on conflict resolution was made on 28 August 2001 by local community leaders, politicians and government officials of Eastern and North Eastern Provinces meeting at that place.

2003, at the height of clan feuds that had claimed about 100 lives. The president declared that a whole community should not be penalized by confiscating their animals for the sins of a few individuals' (*The Standard* Monday 18 July 2005). As a result of this statement from the very top, the then 'Marsabit District Commissioner Muthui Katee dissolved the District Peace Committee and shelved the Madogashe Declaration, alleging that it had no legal backing' (ibid.). Compensation payments are certainly not 'traditional' in Marsabit, where this idea does not figure at all in many types of inter- and intra-ethnic relations.[7] They would have been an innovation. To what extent and with which local modifications they would have been a useful innovation then was no longer discussed.

Also the restitution of animals taken in a raid, with or without additional animals paid as a fine, is not practised by Rendille and Gabra. Here the rule is that 'what has been taken by the spear belongs to the spear', i.e. to who took it by force. In hot pursuit one might recapture animals which are driven away, but once the animals are in the enclosures of the enemy and especially after periodical rituals like *sooriyo* and *almodo* have been performed and the animals have participated in the blessing going out from these ceremonies, they are rightful property of those who took them. In the case of camels, where the distinction between full property (*halal*) and various forms of loan applies, and where the full property of female camels is never transferred, such camels could be only given back as a loan, and a loan from the enemy is not desired. Camels of unclear ownership are believed to bring bad luck and might wipe out entire lineages of those who wrongly claim them. If the government or any intervening agency gave back raided camels to those from whom they were raided, this would therefore be regarded as highly unpropitious by the 'beneficiaries'. In a counter-raid, when Rendille capture camels earlier taken by Gabra from Rendille, these would belong to whoever took them, by the right of the spear, not to whom they originally belonged and whose property marks they might bear. The sons of the latter would rather go for

[7] Rendille and the people they call 'Boranto' (the Gabra and Boran) regard each other as enemies (R. *chii*, B. *nyaab*). Killing an enemy raises one's status. It is not regarded as a wrong that needs to be redressed. In Rendille/'Boranto' relations (like in German/French relations) there has been an alternation of war and peace for generations, without compensation ever having been paid for individual victims (and without the heavy collective 'réparations' imposed on whoever lost the war so characteristic of the German/French case).

In their internal relationships, the Rendille regard each other as brothers. By killing a brother you weaken yourself and thus are punished already. Someone who has killed another Rendille will be symbolically avoided as unclean. No one will borrow his headrest/stool or use any other of his personal belongings or share food with him, until a purification ritual has been carried out (Schlee 1979, 2008a). Exchange requires exchanging units, and to exchange compensation for blood shed requires distinct politico-military groups (in the sense of groups expected to be internally peaceful, potentially hostile to each other, and capable of collective action). A boundary between distinct groups of this kind is simply not given within Rendille society. No matter how intense the rivalry between Rendille clans and moieties might be and irrespective of the number of Rendille killed by other Rendille, such killings are always treated as internal.

In one case, which involved Rendille and Ariaal (rather than 'true', 'proper', 'white' Rendille) and the Administration, on the insistence of the Administration compensation was paid and grudgingly accepted. The animals were first chased away and dispersed and then appropriated somehow, without really making clear by whom (information from Baleisa Hambule). 119

different camels than the ones which once belonged to them, precisely to avoid the impression that anything might have been given back to them. In the mirror case, when Gabra recapture camels from Rendille, the same rules apply. There may be reasons for saying that it is time for these time-honoured beliefs and practices to change, but such changes then cannot take place in the name of 'tradition', a label they often wrongly claim.

Another system of rules with which those of the Madogashe Declaration are at variance is the Law of modern states. President Kibaki was perfectly right in saying that there was no basis in Kenyan Law for penalizing entire communities for the misdeeds of individuals. Modern penal law is based on the concept of individual guilt. But this also applies to British Law, and the British nevertheless, in their colonies, had vast areas in which they abstained from applying their own Law and instead recognized local forms of compensation. This was part of 'indirect rule' which applied to all peoples and regions where direct rule was regarded as too expensive in relation to the advantages it would bring, because resistance would be too great or the returns too little. To go back to such forms of indirect rule would imply the admission that northern Kenya is a special zone in which the Law of Kenya is not applied. Some would say that this would just mean to admit the obvious. When pastoralists massacre each other in a raid in northern Kenya, there will be no prosecutor, no judge and no sentence. In fact, northern Kenya is already treated as a special zone, a zone of tribal warfare.

According to a widely shared opinion, the politicians in the district contributed further to the weakening of the District Peace Committees at different levels: locational, divisional, district and even cross-border peace committees. Politicians are believed to benefit from the hostilities and to incite the two communities. Their influence extends to the Chiefs, whose appointment they influence, and to local civil servants working in the district who have also been accused of leaking and distorting information and fuelling ethnic rivalry.

If we widen the regional perspective to include southern Ethiopia, other factors come into view. The Kenyan Boran have been accused by the Gabra of hosting and being sympathetic to the OLF insurgents. It is alleged that these insurgents have occasionally been used against the Gabra. To date, they still pose a danger to security in the district, although that danger is often exaggerated by politicians accusing each other of supporting the OLF or using them for their purposes. Involvement of OLF in the Turbi Massacre, however, was very unlikely. If the allegation is true that some of the raiders were from Ethiopia, there was no way the OLF, which currently has no foothold in Ethiopia and is hunted down by the Ethiopian government, could raid the Gabra together with any Ethiopian forces. Of recent, the OLF have confined themselves in *fora* (cattle camps) in the Golbo plains. There were cases when they have kidnapped Tabaqa (local militia) near Moiale town. They took them to their *fora* camp, interrogated them and released them without harm, though they took their firearms. The common talk is that OLF have become much friendlier and unlike in the past, they do not kill. Many of them have married among the locals and now live as Kenyans having acquired identity cards to the regret of many people. The Ethiopian authorities once declared in a public meeting that the OLF threat is no longer a concern to them.

Transnational links are also a factor contributing to the conflicts in Marsabit and Moyale Districts. For instance, both the Gabra and the Boran local politicians and community leaders have aligned themselves to rival factions in Ethiopia. This has had a ripple effect in the local politics.

The Gabra in Ethiopia supported the present Ethiopian government under the Ethiopian Peoples Revolutionary Democratic Front (EPRDF), which won[8] the elections in May 2005. It is also alleged by the Gabra that the Boran have always wished that Ethiopia be governed by the Oromo Liberation Front (OLF). The rivalry of the two communities about their standing in the eyes of the Ethiopian government has spilled over to Kenya. Whenever Boran bandits from Ethiopia struck Gabra in Kenya, the Ethiopia government recovered the livestock from the Boran and handed it back to the Gabra. It was alleged that the man in charge of security operations, himself a Gabra Miigo, used his position to favour his people.

Another cross-border issue is poaching. Poaching of wild game too has been on the increase in Kenya. Ethiopian Boran poachers come all the way to Marsabit national park to kill large game. Killing of elephants, rhinos and buffalos is a significant practice among the Boran, and equivalent to killing human enemies. Praise songs are composed for successful raiders and poachers. Songs of scorn and despising are sung for those who are not successful. However, these activities have now taken a new dimension with the introduction of commercial raiding and poaching. Elephant tusks are valued for ornaments in the shape of bracelets (*harbor*) which signify the killer status acquired by killing a man, a buffalo or an elephant. The tusks could be exchanged for cattle. Many Boran youths who travelled to Marsabit and back had to travel through Chalbi, which is part of the Gabra territory. On their return from Marsabit with their booty, many of them were hosted by Gabra and never heard of again. The Gabra were said to kill them in their sleep.

To return from border issues to the highlands of Marsabit: agricultural lands, limited to the high altitudes, are a key resource for the increasing population which can no longer derive its livelihood from pastoralism. Land ownership is hotly contested – not only in the case of the Gabra settlement scheme on Marsabit Mountain. In Songa, for instance, the Rendille were settled by a Christian missionary in 1971 and taught farming skills to reduce dependency on relief handouts. However, the area where they were settled was claimed by the Boran in Badasa as a dry season grazing area. The Rendille there now practice agro-pastoralism while the Boran are primarily pastoralists.

All these settlement schemes were however located on Trust Lands.

[8] The victory is questionable. It was contested and followed by a massive crackdown on the opposition.

The only documentation the occupiers of the settlements can lay claim to are the Temporary Occupation Licences (TOLs). The Boran accused the Gabra and Rendille of using the TOLs for permanent settlement. But in terms of documentation, TOLs are all that people have. To the regret of the Boran, their own older claims are not supported by any documents.

Having so far concentrated on Turbi and Marsabit Mountain, let us now turn to some of the other hotspots of the district.

HURRI HILLS

The Hurri Hills are in Maikona Division and lie on the border with Ethiopia. Their southern end is about 140 kilometres from Marsabit town. It was one of the hot spots of violent conflict in what is now referred to as the conflict belt. The Hills rise 1251 m above sea level. The communities that live here include Boran, Waata, Konso and Gabra. Over the years, there has been an influx of Boran people from Ethiopia.

The immigrants have been issued with Kenyan Identity Cards. This has made the Gabra fearful of being pushed out of the area. Before 2002, elected councillors (for the County Council) of the area have been mainly Boran due to their increasing numbers. In 2002, however, the number of Gabra registered voters also increased and a Gabra was elected as councillor for Hurri Hills ward. Gabra suspect Boran from here of seeking support from Boran MPs from neighbouring districts and from NGOs led by Boran. In reality, this form of politicking might just be the activity of a few individuals and not have a major impact. It does, however, spoil the atmosphere. There is also an element of measuring each other's military by raids and counter-raids. Development activities are adversely affected by all this. NGOs have become victims of misinterpretation of their operations by the warring groups. For example, a tree planting project by an NGO has been at the centre of controversy: afforestation was misconstrued as a deliberate effort to relocate the Boran away from the area.

FOROLE

Forole is about 230 kilometres north-north-west of Marsabit. It is a small trading centre below Mt. Forole at the intersection of tracks from Turbi and Maikona. Though the Gabra predominated, before the conflict the area is occupied both by the Gabra and Boran. This area too, is part of the conflict belt of the district. It is in Forole that the Gabra killed six Boran in their sleep, near the Ethiopian border and looted their livestock. This incident triggered the massacre in Turbi on 12 July 2005.

Forole is a holy mountain for the Galbo phratry (grouping of clans) of the Gabra who hold their age-set promotions there, last in 1986 and in 2000. In such years, they do not tolerate the presence of Boran agriculturalists there (Schlee 1989c, 1990a, 1992d). It remains to be seen in which way the ritual life of Galbo will change if they should be permanently prevented from access to their holy grounds by interethnic strife.

WALDA IDP (INTERNALLY DISPLACED PERSONS) CAMP

Walda is in Uran Division of Moyale District. It borders Rawana which is also in Uran Division; Rawana is the immediate neighbour to Turbi. Both Rawana and Walda were watering points used by both the Gabra and the Boran before the conflict. Although the two communities were initially using resources together, it is now difficult for Boran pastoralists to cross over to Turbi. Similarly, the Gabra have no access to watering points at Rawana and Walda. The number of Boran displaced from Turbi after the incident on 12 July 2005 was 78. Amongst them were 12 primary and two secondary school students. Of these Boran from Turbi, who now stay as IDPs at Walda, several had moved there a few days before the massacre took place in Turbi, which gives rise to speculations that they had been warned by their fellow Boran.

The elders of the Gabra and Boran met just after the massacre in Turbi. The meeting called for ceasefire and for abstaining from further killings, revenge and counter revenge, but to no avail.

From 15 to 18 December 2005 a meeting at Moyale brought together over 150 participants, among them Boran Abba Gada and Gabra *yaa*[9] elders, Government of Kenya officials, Government of Ethiopia officials, and civil society representatives, like those of inter-religious and women's initiatives. This meeting led to adoption of resolutions that were signed by Boran and Gabra community representatives in Kenya and Ethiopia. The government's representatives of Kenya and Ethiopia and civil society representatives witnessed the signing of the resolutions. The resolutions proposed the return of those displaced in Ethiopia and Kenya and to bring back stable peace in the area through 'traditional community' dialogue.

The Boran complaints about the Gabra comprised the severing of ritual ties with the Boran, for instance by abandoning ceremonial sites like the ones at Melbana and Magado. Further complaints were about the application of abusive terms to Boran, killings of individual Boran, and burnings of houses. The Gabra were accused of claiming Boran grazing land, farms, and watering points demanding that the Boran return to their Dirre homeland. In some areas they forcefully took over the Boran traditional wells, like those of Jaldesa in Marsabit, adding insult to injury by asking the Boran to roll these wells up and take them with them. The Boran further resisted their labelling as OLF and accused the Gabra of using this label to turn both the Ethiopian and the Kenyan governments against them. The Boran blamed the deterioration of the Gabra/Boran relationship on the North Horr MP, Dr Bonaya Godana and demanded that he should appear in person to have these matters sorted out.

---

[9] *Yaa* is the term for the mobile capital of a Gabra phratry. Each of the five Gabra phratries (lit. 'drums') has one such *yaa*.

The Gabra from Kenya and Ethiopia had the following grievances. The five Gabra Malbe phratries, *Dibbe Shanaan*, 'the five drums', were not adequately represented in the meeting. The Gabra further mirrored the demand that Dr Bonaya Godana should appear by demanding to see the Boran MPs, namely Hon. Guracha Galgallo and Hon. Abdi Tari Sasura who represented Moyale and Saku (Marsabit Mountain) respectively. They demanded compensation for the killing of women and children, the livestock stolen and the properties destroyed at Turbi.

The meeting called for the cessation of hostilities. The two communities demanded that their traditional authorities and their political representatives be fully involved in the continued peace process. The next meeting was to have been held at Marsabit on 9 January 2006 with the ritual office holders and other representatives of the two communities, MPs in Kenya and Ethiopia, and the Government officials of both Kenya and Ethiopia present. Later the date was postponed to 10 April 2006 where selected delegates from Moyale and Marsabit Districts from the two communities were to meet to discuss possibilities concerning inviting the Ethiopian delegates. It was while people were waiting for the arrival of the delegates from Nairobi that the plane that carried them crashed.

The Ethiopian government had been putting pressure on the Kenya government to bring peace to this region as the conflict in the area was alleged to provide a role to the OLF. The Kenyan government had passed this pressure on to the area Members of Parliament, urging them to resolve the conflict. The two governments brought together elders and local political leaders of the two rival communities, the Gabra and the Boran. The NGOs operating in the area were requested to facilitate transportation and to provide food for the participants in the meeting.

On its approach to the Marsabit airstrip the plane carrying the area Members of Parliament and government officials hit a mountain and crashed. Out of the 17 passengers, 14 died, among them the four Members of Parliament for both Moyale and Marsabit Districts. They were Dr Bonaya Godana, MP North Horr, Dr Guracha Galgallo, MP Moyale, Abdi Taari of Saku and Titus Ngoyoni of Laisamis. Among the dead were also retired Lieutenant General Abdullahi Adan, member of the East African Legislative Assembly, the Rev. William Waqo of the Anglican Church, Assistant Minister for Internal Security Mr Mirugi Kariuki, the Moyale District Commissioner and six police and military personnel. Those killed were all burnt beyond recognition. Only three survived, namely the Provincial Commissioner, Eastern Province, Mr Patrick Osare and two crew members. This tragic event dominated the Kenyan media for days. It had, however, no effect towards ending the rampant conflict in the region.

Three months after the death of the area MPs of Moyale and Marsabit Districts, by-elections were held to replace them. At first it was proposed that the vacant Moyale and Saku seats be occupied by brothers of the departed MPs, with the MPs for North Horr and Laisamis being succeeded by their widows. Both the Gabra and the Rendille were, however, opposed to women's leadership. In Moyale and Saku, campaigning by Boran succeeded in getting enough sympathy votes for the two brothers of the late MPs to be elected. These were Wario Malla and Abdi Tari Sasura. As these by-elections were held just one year before the end of the term, campaigning just continued unabated for the general elections in 2007, in which the new incumbents would have to defend their seats again.

CHANGING ALLIANCES: GABRA WITH BORAN AGAINST GUJI IN ETHIOPIA

The changing alliances which we have examined in detail for the period up to 2000, in recent years have turned up some local facets and undergone some new twists, on which we shall now attempt some updating.

Those Ethiopian Gabra (Miigo) who withdrew from the Boran in early 1990s, lived among the Guji, another Oromo group to the north of the Boran,[10] for several years. They formed an alliance with the Garre and fought against the Boran. The Guji and Boran relationship at the time was stable and peaceful and there was no way the Gabra could convince the Guji to join them in the fight against the Boran. When the war was over, the Gabra continued to live in the Guji territory. The Gabra/Guji relationship declined when the Gabra started to claim part of the Guji territory as their *qebele* (local government unit in Ethiopia). The Gabra also sought administrative positions in neighbouring Guji *qebele*s as well as in the district (*woreda*) offices. The Gabra also demanded their own *woreda* in 2005. The Guji viewed this as an abuse of their generosity as hosts of the Gabra and protectors from the Boran threat. Clandestine murders of Gabra and looting of their livestock continued for several months. Although their relationship with the Boran at the time was trifling, some Gabra families sought refuge among the Boran. The Boran welcomed many Gabra families and settled them in Areero. Realizing that the Boran were no longer hostile to them, more Gabra withdrew from the Guji territory and joined others at Areero. The Boran welcomed them back and settled many of the families at their old settlements in Web. Both the Gabra and the Boran lived at peace and several members of the Gabra elite were employed in the Boran zone government offices. The Guji/Gabra relationship remained hostile.

When the Gabra and the Guji clashed, the Guji retreated to the north of Finchawa abandoning land south of the town. The Boran promptly moved into the area. After the Gabra/Guji conflict subsided, the Guji began drifting back to reclaim their land but met Boran resistance. The Boran and the Guji clashed heavily over several weeks until the government intervened and restored order. A peace meeting was held in Yabello, attended by the Boran and the Guji Abba Gada and state officials. However, though violence ceased, the conflict remained unresolved.

[10] On the Guji and the recent changes in their politico-military alliances see Tadesse Berisso's contribution in Schlee and Watson (eds) 2009.

The introduction of ethnic federalism enhanced the endemic ethnic conflicts in this region. The bone of contention since then has always been where to draw the line to separate the administrative areas controlled by the rival ethnic groups.

The Gabra Miigo are Muslims and have much in common with the Somali. With the Boran they share their language and the fact that they have a *gada* system although, like the different phratries of the Gabra Malbe and the Rendille, not the Boran one but one of their own. They call their *gada* system *dikko*. The Gabra are a prime example of a group between the Oromo and the Somali, being related in different ways to both, without being one or the other. The Boran consider Gabra as Oromo and regard themselves as superior in status to them, as their elder brothers. Traditionally in the past, the Gabra accepted their inferior status within the Boran orbit but lately they have begun to show their resentment. The Somali on the other hand claimed the Gabra as their own. Under Somali influence, a hierarchy of political offices called *talia* has been established among the Gabra (see below), competing with *dikko*. Nevertheless, in the 2004 referendum, many of the Gabra in Somali region voted in favour of Oromia.

Several years after the Gabra had moved into the Guji territory, there were frequent violent clashes and occasional killings between the Gabra and the Guji herdsmen. This was sparked off by the Gabra demand for a separate *woreda* on Guji territory. Another factor that contributed to the escalation of the conflict between the two groups was the Gabra herders' and Guji farmers' conflict over land. The Gabra found it increasingly difficult to maintain their herds because of the expanding *kalo* land enclosures by the Guji cultivators. A series of conflict resolution meetings was held but none has resolved the conflict. There had never been a traditional mechanism for solving conflict between the two groups as they had never before lived together as neighbours.

In March 2006, a Gabra killed a Guji woman and stuck a stick into her private part. The Guji then openly declared war against the Gabra. Over sixty Gabra were reported killed. Hundreds of their houses were burnt down and thousands of their livestock looted. The Boran came to rescue the Gabra and fought against the Guji forcing them to retreat. The Boran settlements at once moved into and occupied areas evacuated by the Gabra and the Guji. Many Gabra fled, many to Moiale and some to Areero. Boran say Gabra have not been grateful for finding refuge there, as many of them subsequently joined their brothers in the Somali region.

In May 2006, a Gabra youth was murdered by a Guji in Areero *qebele*. Boran arrested the culprit and handed him over to the police. A week later, in revenge, Guji forces killed many Boran at Burjuji, a gold mining town in Guji territory, where many people from different ethnic groups lived searching for gold. Many of the Boran victims were mutilated. Limbs, ears and the private parts were cut off. Other people from other ethnic groups were spared. The Boran declared war against the Guji. For five consecutive days, heavy attacks were launched against the Guji. The Gabra joined the Boran forces and led the raiders far into Guji territory. Efforts by the government forces to stop this failed as the Boran also attacked the government forces who were few in number as the southern brigade had already been sent to Somalia at the time.[11] The Boran forces had the advantage and moved through the Guji territory without resistance. Many Guji were killed. In the end the government mediated a cessation of the hostilities. The Boran Abba Gada, Liiban Jaldessa, and the Guji Abba Gada, Aaga T'ant'alo, were mandated to solve the conflict.

FACTORS THAT CONTRIBUTED TO THE WAR BETWEEN THE BORAN AND THE GUJI

A constant feature of recent Boran history is their struggle to contain the Somali expansion. This confrontation began long before the occupation by the Ethiopians and the Europeans at the end of the nineteenth Century. The division between Kenya and Ethiopia also fragmented Boran society and ended Boran hegemony in the region. Ever since, the Boran have been fighting a defensive struggle. In the post 1991 federal arrangement, the Boran lost considerable territory to the east and south-east of their territory to the Somali. Thereafter, they became involved in an often violent territorial dispute with the Garre along the boundary line and for control of Moiale town and district. The 2004 referendum in the disputed areas settled some cases, but not in Moiale where it was not possible to hold it.

The northern part of the Boran territory is partly occupied by the Guji, with whom they share a history of hostility and war. In 2002, the north-eastern Guji-inhabited district of Borana Zone, which had a mixed Borana and Guji population, was hived off to create a Guji zone with its offices at Negelle town, the former capital of the Boran Province. This created great tension between the two groups who had never agreed where the boundary between them should be drawn. Endless violence ensued. Along this border, the Gabra sided with the Boran in fighting the Guji.

GABRA WITH GARRE AGAINST BORAN IN ETHIOPIA

Since 1991, the Gabra have been involved in the many-sided struggle for a share in the Borana zone, i.e. for access to territory there and for administrative recognition. The Gabra are a good illustration of ambivalent ethnic identity, caught between the larger Oromo and Somali nationalities. As the Ethiopians (like the British in the north of colonial Kenya) only recognize those two major categories, the Gabra, who are neither Oromo nor Somali (but a linguistically Oromo-ized people descended from Proto-Rendille-Somali), find a rather awkward choice imposed on them. They are culturally related to both the Oromo and the Somali, having been exposed

[11] On the role of Ethiopia in Somalia see Schlee 2008a.

to older and more recent influences from both groups. Being denied a separate identity, they can find reasons for claiming to be more like the Oromo or for being more like the Somali, and in this they tend to make rather opportunistic choices. Some opted for Oromo and others for Somali. In cultural terms they were forced to choose between Islam and Gada. The Gabra had already abandoned their traditional *yaa* ('mobile capital') institution since the 1974 revolution. At this pont the Gabra living among Boran were forced to revive the *yaa* institution at Webi. On the other hand, the political setup of the Somali region demanded that they set up an institution different from the Gada. The Gabra living among Somali therefore 'revived' the 'long forgotten' institution called *talia*[12] said to be based on the *shari'a*.

The early 2000s were marked by an emergence of the Gabra Miigo struggle for separate recognition and a share in state resources. In Ethiopia, where the federal arrangement provides each ethnic group with their own self governing districts, the Gabra did not acquire this status. In their struggle for a share, they have just made enemies among neighbouring communities. In Ethiopia the Gabra were at war with the Boran, the Garre and the Guji. In Kenya, they were at war with the Boran, the Garre and other Somali pastoralists. In business ventures, they were enviously competing with the Burji hence had a strained relationship with them. The Gabra Malbe too were at war with the Boran, the Rendille, the Desanach and the Turkana over pasture and water and a share in the political representation in Marsabit district. The Gabra often struggle for political representation with the Boran, the Garre and the Burji in both Kenya and Ethiopia. They were fighting on six fronts regardless of their minority status. Realizing that they have lost foothold in support of their struggle, the Gabra Miigo sought alliance with their brothers, the Gabra Malbe in Kenya. Gabra Miigo and Gabra Malbe are aware of being related, as suggested by the shared name and some clans they have in common, but the relationship has never been of any political significance in the past. Having discovered a common mission in the shape of fighting what they perceived as Boran domination, the two Gabra communities managed to form a strong alliance under the leadership of Dr Bonaya Godana, the MP for North Horr. They somehow managed to obtain state recognition and to mobilize state support. The Turbi massacre, for all the tragedy it implied, further boosted support for the Gabra cause.

GARRE/BORAN RELATIONS

Since the 2002 elections, the relationship between the Garre and the Boran around Moyale greatly improved. The Garre had mostly supported Dr Guracha Galgallo, the Boran MP. The Garre were fighting with the Murulle in Mandera district at the time and there was no way they could afford to fight at two fronts. However, having acquired two locations and three

[12] Recent research about *talia* has been carried out by Fekadu Adugna (2009).

wards, the Garre felt content with their share of power within Moyale District. It is also worthwhile mentioning the relationship between the Boran and Garre in Moyale, Kenya, which resonates with the relationship between these two groups in Moiale, Ethiopia across the border. Generally, the relationship between the Boran and the Garre also in Ethiopia remained harmonious in comparison to earlier years. Much political competition is internal to the Garre, who are fighting among themselves along clan lines about power sharing in the Somali regional administration. The line of confrontation between the Boran and the Garre has been the control of Moiale town and district, the gateway to Kenya. Oromo and Somali regional States both claim Moiale town and have set up rival administrations to reinforce their claims. Moiale district belongs both to Liiban Zone in the Somali regional State and an Oromo one that belongs to Borana Zone in Oromia State. The 2004 referendum proved impossible to hold mainly because of the inability of the rival regional administrations to agree who should be eligible to vote. The Somali wanted only Oromo and Somali voters to participate, while the Oromo, counting on the support of a larger number of Gurage, Burji and other immigrant communities, insisted that all residents have the right to vote. Two attempts were made to hold the referendum but both were given up.

AJURAN/BORAN RELATIONS IN KENYA

The Ajuran, whom we found striving for emancipation from the Boran and inclusion in the Somali fold in the early part of the twentieth century and seeking a new alliance with the Boran as soon as that century was over, have managed to remain allies of the Boran, but the relationship has not been free from disruptions. During the Ajuran/Garre war, the Boran fought along with the Ajuran and managed to push the Garre out of Ajuran territory. Had it not been for the Boran, the Garre would have forcefully occupied two locations, to which Garre chiefs were appointed to administer. In the years 2002 to 2006, the Ajuran herders grazed their livestock as far as beyond Sololo. The relationship began to decline when some Ajuran were killed, allegedly by Boran. In revenge, the Ajuran killed two Boran youths and one Sakuye herder. The killers were known and confirmed as Ajuran youths.

On 3 June 2006, Boran elders led by the District Peace Committee held a meeting with the Ajuran in Butte to discuss the unresolved dispute. In previous meetings the Ajuran admitted killing the two Boran youths but denied vehemently killing the Sakuye. The intention of the Ajuran was to admit guilt for killing the two Boran youths and to have the blood price waived in exchange for two of the murdered Ajuran, whom the Boran had admitted having killed. They strongly wanted to put the case of the killing of the Sakuye youth aside and urged the Boran to leave the matter to the Sakuye themselves. They tried to convince the Boran elders not to advocate for the Sakuye. But the Boran elders reminded the 129

Ajuran that 'the Sakuye too are Boran' and that the three cases should be treated alike. When the Boran elders realized that the Ajuran intend to isolate the Sakuye, they gave their verdict in the following words.

> Now time has come for you to separate from us. Leave our land, our pasture and our water. People who have grudges about each other never live together. We have provided you with much hospitality. In the past you have denied us your water and pasture. Now it is my turn to act in such a way. Know that if you resist leaving our land, you too will be moved out in the same way that you moved us out of your land.

An Ajuran elder desperately remarked:

> Is it practicable that we two can live without each other?

A Boran elder replied:

> Yes, of course, and if we ever move into your land deny us your pasture and water. It is you who need us and yet you cannot admit to an offence yourselves committed on our land. You deny us the *aada* that we have in common. Now that you have failed to honour the *aada*, let us separate.

The sentiments expressed by the Boran have had an immense impact on the future of the Ajuran/Boran relationship. For the Ajuran there was no acceptable option. To admit guilt would mean a disgrace as far as their attitudes towards the Sakuye were concerned. To refuse to admit guilt would mean they will no longer have opportunity to graze in Boran land. So the Ajuran pleaded with the Boran to give them time to consult with their community and arrange for another meeting.

Not long after this, a relative of one of the Boran victims organized a raid on the camels of one of the suspected Ajuran killers. This raid did not cause any alarm as the Ajuran themselves had previously blamed the herd owner for causing a rift between them and the Boran.

Several months later, the Ajuran raided Boran cattle and made away with over one thousand heads of cattle. The newly elected MP for Moyale, Hon. Wario Malla, the brother of the late MP Dr Guracha Galgallo, made consultations with the local leaders, the provincial administration and the MP for Wajir North, Dr Ali Abdullahi and the return of the looted Boran cattle was negotiated with success. The MP, Dr Ali, made all efforts to effect the return of the looted stock and the animals were handed over to the Boran leaders. This was witnessed by the government representatives of both Moyale and Wajir Districts. However, the Ajuran also demanded for the return of the looted camels, which the Boran leaders promised to return. Though many attempts were made to ensure the return of the Ajuran stock, there seems to be little hope of this ever materializing. This however, has not strained the relationship between the two communities to an extent which would bring it anywhere near the breaking point. The Ajuran are not in a position to push this issue too much, and therefore it may remain unresolved.

A RECURRENT FEATURE

A recurrent feature we have encountered in these brief histories of recent conflicts is failed peace meetings, which either remained inconclusive or whose conclusions were not implemented. When Schlee, in 2007, asked a Gabra interlocutor at Bubisa, a location in the lowland immediately to the north of the Marsabit Highlands, why this was so, the answer was: 'Because of the townspeople.' It is pastoralists who fight and are killed, but it is the townspeople who are behind the conflicts. Also the boundary issues are heated up by townspeople. If it was just the pastoralists and their resource allocations, they would manage these matters as they have always done.

He illustrated this position taking the Boran of Marsabit, at that time the deadly enemies of the Gabra, as an example. The townspeople and especially their political leaders want a separate district[13] to be shaped in such a way that the Boran would have a secure majority there.[14] The Boran pastoralists, however, would prefer a district which comprises the lowlands inhabited by the Gabra, so that they can make arrangements with the Gabra about the seasonal use of these of these pastures, rather than being confined to their parts of Marsabit Mountain, a mere island in the lowlands, where pastureland is scarce anyhow and further limited by the National Park and by the expansion of agricultural lands.[15]

Such conjectures by Gabra about the motivations of their Boran opponents, distinguishing between different types of Boran opponents, appear quite plausible. One may try to generalize this type of reasoning. Across the border, in Ethiopia, and in many other parts of the world, where little nations are carved out of larger ones, or provinces or districts subdivided to give special rights to special people, this may well be in the interest of elites who rely on ethnic constituencies and sedentary people who want strangers to quit their farmlands or business areas. It hurts, however, the interest of mobile occupational groups like pastoralists and traders, and also those of ordinary mixed farmers who practice transhumance with some of their cattle. It also kills what is most human in us, namely our ability to communicate with each other in spite of our differences or even through these differences, to learn each other's languages, and to find arrangements to get along with each other. Ethnic purity does away with much of the challenge to bridge differences. There is a narrow link between purity and boredom.

On August 4, 2010, Kenyans held a referendum on a new Constitution.

---

[13] In 2007 the Government planned to create several new districts all over the country. It was proposed to divide Marsabit District in three parts: Marsabit North, Marsabit Central, and Marsabit South. The opposition suspected electoral politics behind these moves.

[14] In such contexts one might speak of 'ethnic un-mixing' or even of 'negative conquest', if the aim is to shed ethnically mixed areas to create ethnically homogeneous administrative units and electoral constituencies.

[15] About resource conflicts on Marsabit Mountain see Adano and Witsenburg 2008.

There was a clear majority of 'yes' votes.[16] Among other things the constitution provides for a new regional order. After a number of splits which have produced smaller districts, these are now combined into major units, called counties. In our area of investigation, there are two such counties, Marsabit and Isiolo, which form part of a new region called Upper Eastern. Marsabit County comprises Marsabit, Moyale, Sololo, Chalbi, Laisamis and Loyangalani Districts and Isiolo County comprises the Districts of Isiolo, Merti and Garba Tula.

The new constitution has some clear legal provisions that shall directly benefit the people of the Upper Eastern Region. The pastoralists and other marginalized communities who have experienced abuses, violation of their human rights and persecution under the old constitution, will gain benefits in the new constitution under many provisions especially in chapter 3 'Citizenship', chapter 4 'The Bills of Rights on Equality, Rights and Fundamental Freedoms for all Citizens', chapter 5 'Land and Environment', Chapter 7 'Representation of the People', chapter 12 'Public Finance' and chapter 16 'General Provisions'.

Following a global trend to recognize collective identities and group rights below the level of nationality, the new laws establish the category of 'marginalized communities' as deserving special attention. In Article 260, a marginalized community is identified as:

a) A community that, because of its relative small population or for any other reason, have been unable to fully participate in the integrated social and economic life of Kenya as a whole.

b) A traditional community that out of a need or desire to preserve its unique culture and identity from assimilation have remained outside the integral social and economic life of Kenya as whole.

c) An indigenous[17] community that has retained and maintained a traditional lifestyle and livelihood based on hunter and gatherer economy or

d) Pastoral persons and communities whether they are:

    (i) Nomadic or

    (ii) A settled community that, because of its relative geographic isolation has experienced only marginal participation in the integrated social and economic life of Kenya as a whole.

The term 'marginalized groups' refers to groups of people who because

---

[16] Among the Rendille there seems to have been a considerable proportion of 'no' votes. The reasons for this have nothing to do with wider politics but with Rendille family law. The constitution defines rights of extramarital children against their genitors. In Rendille customary law, genitors are completely ignored, although who has actually begotten whom may be a matter of rumours or in certain cases even general knowledge. Rendille children belong to the husband of their mother, and the bridewealth has been the payment for the establishment of paternal rights (Schlee 2009b). Rendille fear that the whole fabric of responsibilities among them would be dissolved if men had to care for children they have begotten elsewhere rather than building up the herd for their own sons and heirs, legitimate in their terms, and if these latter turned to real or suspected genitors elsewhere for support and cooperation rather than submitting to paternal authority and contributing to the family economy.

[17] We shall address some problems in applying this category to Africa below, in Chapter 3.

of laws or practices before, on, or after the effective date were or are disadvantaged by discrimination on one or more of the specific grounds laid down in article 27 (4). According to Article 21 (3) all State organs and public officers have the duty to address the needs of vulnerable groups within the society, members of minority groups or marginalized communities, and members of particular ethnic, religious or cultural communities.

The new constitution provides undertaking for equal treatment of all citizens, enjoyment of rights and freedoms and discrimination of any form that is based on ethnicity, religion, race, or beliefs. From the chapter on citizenship, the Bill of rights from article 12(1), the new constitution is expected to provide new gains for the people of Upper Eastern Region on various issues that include equality, freedom, secure land administration, property protection and security.

A precaution is taken that the demand for equality cannot lead to the reduction of existing rights of a certain category of women, namely Muslim women. Article 24 (4) protects the rights already enjoyed by Muslim women. These cannot be taken away by the new constitution. These rights enjoyed by Muslim women under the Muslim law include the following: the right to own property, the right to inherit from father, mother, brother, sister, husband, daughters and sons, the right to be executor, administrator, trustee and beneficiary (of *Wakf* – land grants), the right to consent to marriage, the right to be provided fully by their husbands in terms of shelter, food, clothing, medical and all her maintenance and upkeep, the right to have her children provided for fully by the husband, the right to fully retain what she owns and/or earns during marriage, the right to seek divorce whether on grounds or not, the right to have custody and maintenance during *eddah* [the waiting period] after divorce, the right to shelter and maintenance during *eddah* after the death of husband and the right to be paid for breastfeeding her child/children.

Article 63 recognizes the ownership of community land. The new legal framework legitimizes authorities of the communities on the basis of ethnicity, culture, or similar community of interest to manage their land. The parliament is obliged to enact new laws that shall recognize and encourage the application of traditional dispute resolution mechanisms and task the government to initiate investigations, on its own initiative or on complainant, into present or historical land injustices, and recommend appropriate redress. These provisions are captured under articles 67(2) (e, f) and article 159 2(c).

The new constitution breaks away from the past traditions of not paying attention to incidences where many neighbouring districts are in conflict with each other over boundary disputes: Article 188 provides the conditions for altering the county boundaries.

The culture and language of the people are protected by the new constitution. According to article 44 (1) every person has the right to use 133 the language, and to participate in the cultural life of his or her choice. Article 44 (2) provides that a person belonging to cultural or linguistic community has the right, with other members of that community (a) to enjoy the person's culture and use a person's language or (b) to form, join and maintain cultural and linguistic association and other organs of civil society and (3) a person shall not compel another person to perform, observe or undergo any cultural practice or rite.

Article 170 (1) retained the Kadhi's court in the constitution, though as subordinate court against article 2 (6).

Pastoralists and marginalized groups are accorded control of the counties in article 174 (a) (b) (c) (d) (e) (f) (g) (h) (i) and gain more voice and representation in the National Government through the Senate, and in the Cabinet and employment.

Article 91 requires all political parties to respect the political rights of everybody including minorities and the marginalized groups under clause (a) and (e). Article 100 further compels parliament to enact legislation which will promote the representation in parliament of a) women, b) persons with disabilities, c) youths, d) ethnic and other minorities and (e) marginalized communities.

Article 177 compels the parliament to enact laws that will increase the number of members of marginalized groups including persons with disabilities and youth from the regions in the decision making organs of state at both levels of government.

The new constitution guarantees the marginalized communities some affirmative action programmes that shall be designed to ensure that minorities and marginalized groups:

a) participate and are represented in governance and other spheres of life;
b) are provided special opportunities in education and economic fields;
c) are provided special opportunities for access to employment;
d) develop their cultural values, languages and practices; and
e) have reasonable access to water, health services and infrastructures development.

Affirmative action is defined to include any measure designed to overcome or ameliorate an inequity or the systematic denial or infringement of a right or fundamental freedom (Committee of Experts 2010; Kenya Human Rights Commission 2008; Shongolo 2009).

It needs, however, to be noted that the recognition of communal land rights remains very general and is void of legal instruments. The communities have no title deeds. The new constitution is silent about vast community lands put under the national parks and national reserves. The yet more complicated question of multiple and overlapping land rights of different pastoral groups, so characteristic of northern Kenya, has been left out completely. So communal land rights are neither effectively protected against land appropriation by the State nor against privatization, and they are not defined in their relationship to communal rights of access by other groups than the titular or predominant one in a given area or district.

We can therefore only come to a similar conclusion to the one we reached about the AU Policy Framework for Pastoralism in the introduction, namely that the new constitution is a document which provides hope for a better future for the pastoralists but not the certainty of it. What matters will be the implementation.

3

Feedback and Cross-fertilization: The 'Declaration of Indigenous Communities of Moyale District'
GÜNTHER SCHLEE

On Wednesday, 12 April 2000, a group of thirty-two signatories issued a declaration at Moyale, entitled the 'Declaration of Indigenous Communities of Moyale District'. It was distributed to the District Commissioner, other Government officers, and representatives of the Garre community. These last perceived it, not without cause, as a notice to leave. It was sent to the Office of the President and the Provincial Commissioner. Its contents appear also in another letter to the Provincial Commissioner and other documents. Abdullahi Shongolo was present when it was formulated and even helped in brushing up the English of the document with mixed success, as we shall see. He is therefore in one of the classic role conflicts to which participatory observation leads: how can one analyse or critically evaluate that to which one has contributed oneself? That might not be unfeasible, but it is complicated. He has preferred to leave that task to me, himself being around to answer questions. A closer reading of the text, however, reveals, that I myself have unwittingly and unintentionally also had some influence on it.

Habent sua fata libelli – 'Also books have their fates', the ancient Romans used to say. This saying was confirmed by a new interest in my book *Identities on the Move* (Schlee 1989a) in the area, which now seemed to have been appropriated for the Boran cause. Hitherto, I suppose because of the fact that my wife is Somali, I had been suspected of pro-Somali leanings, or so I thought until I developed the habit to conduct an electronic search for my name from time to time to find out what people think about me. In the internet, I found myself accused of much worse crimes than sympathy with individual Somalis.

In 2007 I came across an internet publication accusing me of 'monumental ideological myopia' or, alternatively, having accepted 'a big bank cheque', being a 'Neo-Nazi', of 'criminal character', stealing ideas from several proposals sent to my institute, etc. The crimes I was identified with were to identify some ethnic movements as narrow and aggressive, not acknowledging that these were based on differences which date back 'millennia'. Like other 'bogus academics' I was accused of favouring 'Semitic focused research' and failing to notice 'how lower Axumite and Abyssinian civilization has been, if compared to the diachronic Cushitic culture that permeated civilization from Ancient Kush and Meroe down to 137 modern Oromos, Sidamas and others'. This anti-Cushitic and pro-Semitic bias has purportedly led me to 'rejecting Oromo, Afar, Sidama and Ogadeni postgraduate students and welcoming Amhara and Tigray clannish elite graduates'.

It so happens that I have never compared ancient Axumite and Meroe civilizations and if I ever did so, I would certainly refrain from denouncing one of them as inferior to the other, as the writer of this internet article appears to want me to do. It also happens that we have numerous Oromo doctoral and postdoctoral students and only one Amhara who however is a protestant and not of a 'clannish elite' origin. I hasten to add that this is not due to a bias opposite to the one I am accused of, namely an anti-Semitic bias (in the sense of anti-Amhara, anti-Tigray or any other sense for that matter) but to the simple fact that I have spent my working life studying Cushitic cultures and languages, and that that is where my research interests and my linguistic qualifications are.

So I have not only been accused of being pro-Oromo or pro-Somali but also – worst of all – pro-Amhara. The permissiveness of the internet allows someone to proclaim all this in public without the least regard for facts. The lunatic fringe of different movements and quite opposed ideological persuasions all accuse me of siding with the others and not with them. My attempts to be guided by empirical facts and not by biases and sympathies have resulted in rejections by all of them. At least that is a quite balanced picture.

The decline of the standards of debate is not only driven by the permissiveness of the internet but also by the proliferation of academic institutions which compete for students rather than for scholarly standing. Another finding in the internet from 2007 was an MA thesis from 2006 by an Oromo student from a 'Norwegian University of life Sciences, Department of International Environment and Development Studies' in which I was accused of calling the peoples living under the Boran hegemony 'vassals' of the Boran, and even of accusing the Boran of 'slavery' and 'oppression', of course without any textual evidence, because such statements cannot be found in my writings. Such unsubstantiated statements should not have appeared in a carefully supervised thesis. Even being wrong, however, such accusations in the present situation of interethnic violence and lawlessness may well put someone at risk of life and limb.

Apart from taking up themes from *Identities on the Move* (this time with a pro-Oromo twist), also another text may have found an echo in this 'Declaration'. The article 'Local War and its Impact on Ethnic and Religious Identification in Southern Ethiopia', by Abdullahi Shongolo and myself (included in Schlee with Shongolo 2012), circulated in the area in a version published in 1995. In the text of the 'Declaration' it is difficult to disentangle selective readings of our own texts from other sources of inspiration. But I will try to point out some obvious borrowing. I will also try to link the text to other discourses, more influential than anthropological writings.

The most important discourse from outside, the godfather after whom the child is named, so to say, is already betrayed by the word 'indigenous' in the title of the Declaration. United Nations programmes, Non-Governmental Organizations, Environmentalist and Human Rights groups have advocated the rights of 'indigenous' populations increasingly over the past years. The UN even declared the decade from 1995 to 2004 the 'Decade of Indigenous People'. Depicting oneself as 'indigenous' is a way of appealing to outside solidarity. It is, however, far from clear, what exactly is meant by this term.[1] In the cases of the Americas, Australia, and Siberia, the meaning of the term is pretty obvious. Here the earlier occupants have been largely exterminated or otherwise replaced by settlers of European origin, and the term 'indigenous' refers to the disadvantaged minorities who descend from these earlier occupants. But how about Africa, a continent whose population as a whole has never been replaced by immigrant outsiders who gained numerical predominance? Are all people, or at least all sufficiently dark-skinned people, now living in Africa, therefore 'indigenous'? Or does one need to prove 'historical continuity' to some pre-colonial entity which no longer exists and whose remnant population lives under the dominance of others, as some definitions suggest, to qualify as 'indigenous'? Does that not open the door to all sorts of fabrications? Is 'consciousness' a criterion, i.e. does one become 'indigenous' by claiming it? How about the interwoven histories of the ethnic groups of northern Kenya and their interethnic clan relationships? Is anyone 'indigenous' who can point to long residence in the area of a group to whom he or she is linked through any ancestor or ancestress? Or does the UN endorse patrilineal descent reckoning or any other descent rule which predominates in a given area? If minorities who cling to their old settlement areas are thereby classified as 'indigenous', does that imply that they deserve more protection than minorities which have been expelled from their older abodes and now live elsewhere, not as indigenous people but as immigrant strangers?

The irony in the Moyale case is that the Boran depicted themselves as 'indigenous', although they, the lords of the land for centuries, clearly had no minority status. And they claimed this status against the Garre, who partly descend from bearers of the Proto-Rendille-Somali culture who might have lived in the area long before the Boran who probably expanded into what now is Moyale District only in the sixteenth century. But the only time period to which reference is made in the document is the colonial period:

Declaration of the Indigenous Communities of Moyale District
Memorandum
A. Historical Precedent:

It is noted that during the colonial era no single member of the Garri community was settled in Moyale except the family of Issack Mohamed

---

[1] This lack of precision has also been remarked in Kenya. In the *Daily Nation*, 11 April 2000, an article was published entitled 'Confusion persists over meaning of "indigenous"' by Chander Mehra.

Guutu and probably several soldiers who were working in Moyale. Yet it was not until the formation of political parties such as the NPPP (Northern Peoples Progressive Party) in early 1960s **when the Garris claimed the Northern Frontier Districts of Kenya on behalf of the Somali government.** The first influx of the Garris into the interior northern towns of Kenya started at that time. A few of them came to settle in Moyale. Notably their presence was again witnessed after Kenya attained independence, when **Hassan Gooro,** and **Alio Gababa** the Garri warlords, launched guerrilla attacks on Kenyan nationals mainly the Boran and Gabra dominated areas of Moyale, Marsabit and Isiolo districts. They killed many Boran and Gabra, looted thousands of their livestock and took hundreds of families captive back to Somalia. [Emphasis in the original]

'Warlord' is an anachronism. Originally referring to China, the term was applied in the African context mainly since the fall of the Siad Barre regime in Somalia. To apply the term to *shifta* leaders of the 1960s makes these participate in the negative image of the Somali warlords in the Mogadishu-Kismayo area in the 1990s. That is, of course, what the choice of this word is meant to convey.

After the demise of the Shifta war, all communities were repatriated to their respective home districts from Somalia. All the Garris were taken to Mandera district. The Garri emergence in Moyale begun yet again with the Mangistu Regime coming to power in Ethiopia in 1975. They settled in many parts of Moyale and other border towns in Ethiopia. After the fall of the Siad Barre government many Garri fled and settled in the eastern parts of Moyale district.

In 1977 the Garris forced out the indigenous communities of Nana and Godoma by means of clandestine murders and other ill-fated means. It was not long before they started trouble when they destroyed an Army lorry and killed many Army personnel in that area. When a major operation was launched against them, they migrated back to parts of Mandera and Wajir districts but re-emerged back in 1986, when the Moyale 'Gold Rush' was excavated. The influx saw many Garris settling in large numbers in Moyale and especially in the town area.

In 1986 the word of a 'mountain of gold' which had been discovered north of Moyale in Ethiopia spread like a bush fire. A shanty town sprung up there. Small scale digging by thousands of individuals, 'informal' mining, so to say, lasted a couple of years.

B. Background History to Territorial Claim

After the fall of Mengistu, several ethnically based political movements came to this region and were struggling for political status. The major groups were the OALF (Oromo Abbo Liberation Front), which was predominantly composed of the Garre community members. Hassan Gooro, the former Somali government general and the one who led the shifta war of 1967, arrived to help the Garre gain occupation of Boranland and for the Garre to dominate the position in the newly established government. The Garre, in trying to form an alliance with the Gabra in territorial claim, made great efforts to create enmity between the Gabra and the Boran. Some of the Gabra youths were forcefully recruited into their army. In an attempt to cause a major conflict between the Boran and the Gabra, the Garre warlords carried out secret clandestine murders among the Boran and the Gabra. Ethnic conflicts flared up to the advantage of the Garre community.

As for those who were fighting, the war was about pastureland and wells. But according to the Garre officials and their warlords, it was a struggle for position both in political and administrative sphere of the government. Their party, the **Oromo Abbo Liberation Front**, which was formerly Somali Abbo Liberation Front, took le[a]d to destabilise the peace of this region.

The beginning of this paragraph seems inspired by our 1995 paper 'Local War and Its Impact on Ethnic and Religious Identification in Southern Ethiopia', where we state: 'For the combatants themselves the war was about wells and grazing areas ...' The text goes on to explain that the OALF had been fighting on the Somali side during the Ogadeen war of 1978 against Ethiopia. Now they were struggling for control of the Boran Province of the Oromo state of Ethiopia. The document claims that at the same time the Garre want to incorporate Moyale into Region 5, the Somali region, which they anticipated to dominate. The Garre had always had the dream to control part of eastern Ethiopia. Now, the document claims, it is clear from this that their next target would be to incorporate parts of Northern Kenya and particularly Moyale District, as the situation is now, into Region 5 of Ogadeen of Ethiopia. There is no doubt therefore that, as history proves, the Garre are motivated by an expansionist ideology, notwithstanding menaces and atrocities rooted deep in their culture of territorial claims.

As to the 'expansionist ideology' of the Garre, I am afraid that my comparison of the self-restrictive adaptation strategy of the Rendille with the expansion strategy of the Somali in general, not particularly the Garre, in *Identities on the Move* (Schlee 1989a pp. 47–51) might have been consulted. Possibly also copies of my 1988b paper on 'Camel Management Strategies and Attitudes towards Camels in the Horn', where these matters are treated more fully, were circulating in the area. Those parts of my publications which are critical of the strategies of ethnic exclusivity advocated by the authors of the Declaration, have, of course, not been adduced.

Statements of the current Problems and Issues:

We the indigenous communities of Moyale district have noted with much concern the recent atrocities and unbecoming tribal clashes perpetuated by the Garre community that is tantamount to be the cause of insecurity and civil unrest in Moyale district. Despite our long patience, we are therefore, obliged to state the following as events leading to the Garre community's causes of civil arrest [unrest] in this district: [. . .]

What follows here is an enumeration of seven incidents. In six of them Garre have killed Boran or Ajuran, in one of them the case of a Garre man who was overpowered by a woman and arrested with three hand grenades is included because of its comical element, although no one was killed. The remaining incident is the explosion of a bomb in Ethiopia which the Garre are said to claim to have taken place in Kenya with the aim of 'tarnishing and derogating the Republic of Kenya and the Kenyans at large.' Then 141 follows the account of the arrest of two Garre elders on 4 April 2000, which has been described above in chapter 1 on 'Moi era Politics'.

7. [. . .] Following their arrest, the Garre community both from Kenya and Ethiopia besieged the District Commissioner's office and the police station in Moyale town. The demonstrators armed with stones, sticks and other arms, protested at the arbitrary arrest of their two community leaders and were demanding for their release without condition. They pelted stones at the D.C's office and directed insults at the local D.C. who only acted within the legal framework to arrest and interrogate the Garre elders who were allegedly involved in fuelling ethnic violence in the area. The Garre community who numbered to about two thousand men, women and youths behaved in an unruly manner and staged an illegal demonstration against the government. The two community leaders, whom they referred to as the leading warlords, were known in the past to have spearheaded ethnic clashes in the area. However while marching along during the demonstrations, they sung provoking slogans and chants in praise of their territorial claim and asking other 'Somalis' to join them.

8. The two-day demonstration, caused chaos and both government and public business came to a standstill. Notwithstanding the fact that the rowdy group tore down the Kenyan National flag at the chief's office and instead hoisted an Ogadeen Provincial Political Party flag, in view of launching a claim on parts of the Sovereign Kenyan soil, was a clear indication that the Garre community had far reaching ambition to violate a national territorial boundary, aimed at causing yet an anticipated conflict between Kenya and Ethiopia. Consequently, in a further incitement to provoke the indigenous community of the area, a member of the Garre community from Moyale, announced over the BBC radio that 90% of Moyale district residents composed of the Garre majority. This is clear evidence that the Garre claim of Moyale district was evident. A map of territorial claim has already been published and is already circulating among the Garre elites, where Boran and Gabra communities of Moyale district were not shown as represented on the map enclosed.

This map has been discussed above (see Chapter 1). It does not contain any territorial claim as alleged here.

9. That in the face of all the provocation, the indigenous community remained tolerant and watched the events with patience. In response to the Garre community's illegal demonstration against the government, the indigenous communities staged a peaceful demonstration to pledge loyalty to the government and to refute territorial claims over parts of Kenyan soil. But the Garre youths threatened them and threw stones at the peaceful group. This was followed by two days of tension with all shops remaining closed.

The Garre community had renewed their old hatred for the Gabra and the Boran communities. They attacked the Boran and Gabra and the government security using dangerous weapons such as guns, clubs and stones. They as well looted many of the shops belonging to the Gabra, the Boran and the Ajuran.

10. In a show of their might and financial power, the Garre brought to Moyale a journalist, a biased Human Rights group, a lawyer and an opposition politician, and a one time Parliamentary aspirant from the North-eastern, himself a Garre, to show proof of their control of what

they now allege to be an area in their control. This was not short of a continued process of their power pride and influence over the government's authority. They had earlier directed accusation to the local D.C. over the arrest of the two community leaders. Using their political and financial power, they immediately influenced the transfer of the local District Commissioner and thereafter a District Commissioner of their choice was flown in a chartered aircraft that brought their lawyer and the pressmen. The Ford Kenya Politician [. . .] who had been frequenting Moyale district for the last several days to serve the interest of his Garre community, escorted the in-coming D.C. on the plane.

That the change in the office of the DC was in any way influenced by the Garre was denied by Government. The Provincial Commissioner expressed herself very clearly on this matter shortly after at a meeting in Moyale.

11. That the trends of the Garre expansionism are not exceptional to Moyale district only. The fact is the problem rooted by the Garres is experienced beforehand in the neighbouring Wajir and Isiolo district. At this juncture we should mention that:

I) The Garres have already claimed parts of Danaba and Irees T'eeno locations where a chief and an assistant were appointed. Subsequently war erupted between the Garres and the Ajuran where the later community acted to regain their land back.

II) Inevitable clash was evident at this location and Godoma location in Moyale district, as a result casualties and deaths filled the air.

III) Evidence in view of the above can be seen in the light of the presence of the Garre MP from Mandera West, Hon. Amin in Moyale. The MP was forced by deliberations of peace meeting in Butte Sub–District to seek salutations from the 16 warlords and perpetrators of the war residing in Moyale. The 16 warlords[2] are short of nothing but planning wars of expansionism against the Ajuran and Moyale district. He had been organising secret meetings with only the Garre community.

IV) And in preparation of an eminent war the Garres are at the moment shifting all their movable properties, livestock, children and families across the border to Ethiopia.

Finally it is beyond reasonable doubt that the Garres at large have acted in a manner, to forcefully acquire our district, perpetrated war to cause insecurity, jeopardise our constitutional rights of enjoying the tenets and fruits of our Nation and deprive us of our nature survival rights. They have created a state of violence with the aim to disrupting peace and stability in Moyale district.

Therefore we hold the following resolutions:

D. Resolutions.

That we the indigenous community of Moyale District do hereby pass these resolutions as an obligation and commitment of peace, survival and rights of sustaining all that appertains to our District that:

1. The presence of Garres in Moyale District has already caused poverty, insecurity and enough bloodshed, thereby their continued presences will

[2] What is meant by this is that the MP had to get the consent (not *salutations*) of 16 Garre businessmen from Moyale (here: *warlords*) without whom a peace agreement would not be worth anything.

only further aggravate the situation and continue to cause more blood-shed and deprive us the right to enjoy our natural environment and peace.
2. Where by that is stipulated here in we demand that the Garre community should vacate and absolutely from what entails the Jurisdiction of Moyale District within the next three weeks w.e.f. 13th April 2000, without failure whatsoever.
3. Thereupon we hold that our Government shall act in accordance with our demands and facilitate a prompt removal of the Garre community from Moyale District. They should migrate back to wherever they came from.

32 signatures, comprising those of Boran, Gabra, Sakuye, and Burji elders.

That the diction of this document is one-sided and polemical is obvious and does not need to be shown by examples. The entire text can be taken as such an example. We reproduce it here to show that the territorial logic of the nation-state with rightful residents and intruders has been fully transferred to the district level by the authors of this document.

In a letter handed over to the Provincial Commissioner on her visit to Moyale, it is even clearer than from the above document, that people of a certain level of education and a certain reading background are behind the elders who sign such documents. A 'culturalist' diction is very obvious:

[. . .] Garre social and cultural relations with the host community were characterised by a widespread antipathy. They have employed all means to antagonize the host community using aggressive means, which they inherited from the Somali culture. Integration of the alien Garre communi-ties into the mainstream of the host communities has been one of strife and bitter hostilities. The recent Garre migration into this district gave rise to a situation in which they suffered from a negative social image that associated them with corruption, banditry activities, robbery, religious fundamentalism, illegal trade transactions and as well many other social ills, thereby influencing the dominant culture of the host community. [. . .]

A sentence phrased as if it was meant to sympathize with the Garre who suffered from a bad image, but obviously meant to alarm the Government about social evils like banditry and fundamentalism!

[. . .] Garre culture [. . .] does not conform to the traditional culture of the indigenous community of this district. They have already polluted the social norms of the host community members. Hence this anti-social practices have posed a situation of major hostilities and would continue to do so, now that the indigenous community would no longer tolerate.

Pollution as a topos of nationalist rhetoric and predictions of violence as covered threats are familiar from the Balkans and the rest of the world. It is a pity that northern Kenya, where almost everyone is related to every-one else, could not escape these divisive ideologies.

4

Some Comparative Perspectives, Conclusions and Recommendations
GÜNTHER SCHLEE

'In 1980 Isiolo District was flooded by pastoral Somali from Wajir and Mandera districts who were fleeing from constant harassment by their fellow Somali, who roamed about, robbing and raping at gunpoint' (Schlee 1989a p. 52). The problem is older than the breakdown of the Somali state in the late 1980s which led to the dictator Siad Barre having to flee the country in 1991 and the subsequent faction fighting and mass emigration. It is older even than 1980, when I made this observation about the state of affairs in Isiolo District.

The recommendation of Boran politicians derived from such states of affairs is forced repatriation of Somali to their presumed districts of origin. They claim the districts in which Boran predominate numerically to belong to the Boran. They cannot even be blamed for this. Murder, arson and looting by the politically dominant 'Kalenjin', which led to the expulsion of immigrants originating from other parts of the country from Rift Valley Province in the early 1990s, have set the precedent. If other 'tribes' claim to have privileged access to the resources of what they perceive as their territories, why should the Boran be the last ones to do so? All the more so as in neighbouring Ethiopia ethnic territoriality is not only the actual practice but written into the constitution.

But is that the only possible conclusion? Let us go back to 1980. (Many other points of time could be used to illustrate the same argument.) If in the Wajir and Mandera Districts the girls had been in a position to take their rapists to court, if the police there had gone after rustlers and recovered the raided stock, if illegal guns had been confiscated, in other words, if there had been a functioning government with disciplined armed forces, no mass migrations from those areas might have occurred in the first place. This argument might sound entirely fictional to Somali from northern Kenya. Their only experience of the 'Government' often is that of just one more gang of armed men who rob, rape, torture and castrate to the pleasure of their hearts. But fictional as it may sound in this regional context, this argument is derived from a wide-spread set of ideas about what a modern nation-state should be like. And maybe these ideas should not be given up too easily.

The basic problem of the northern lowlands is the same as that of the rest of Kenya and of many other African countries. State institutions, 145 including the law, no longer command loyalty. The political class themselves have come to regard the state as a source of private appropriation. There is a spiral of corruption: everyone makes sure that he does not steal less than his competitor. Politicians are expected to receive kick-backs, to misuse public resources, and to distribute a part of their loot among their co-ethnics. They are blamed if they do not do 'enough for their own people'. I do not know how to reverse this spiral. But let us suppose it could be done, let us suppose there could be a new start somehow. Suppose policies could be formulated, which would actually be implemented and not converted into money by officials who receive payments for tolerating their circumvention – which policies would we then wish to formulate for the pastoral areas of northern Kenya and the Horn of Africa?

Loyalty needs to be paid for. Even kings 'by the grace of God' had to give something in return for the loyalty of their subjects: they had to protect them, speak justice, and strike coins of stable value to facilitate their transactions with each other. The same applies to institutions on all levels, including that of the state. If you wish people to feel and act as Kenyans or Ethiopians, participation in Kenyan or Ethiopian institutions needs to pay out for them. The key institution of integration into a modern nation-state is the school. In 1984, in El Das, Wajir District, no geometry was taught because too few parents had been able to buy the mathematical set for their children at KSh 40. If pupils from that primary school wished to go on to secondary school, they would lack the points from a whole sub-discipline of mathematics. This is just one example I happened to find in my field notes. I am sure examples from the quarter of a century which has elapsed since are myriad. The situation does not seem to improve. The quality of teaching in northern Kenya leads to joblessness and criminality in Nairobi and elsewhere. But even those who by miraculous circumstances acquire good school leavers' certificates can no longer regard them as entry tickets to employed work. Education no longer holds the promise of integration.

School leavers often cannot even be re-integrated into the local livestock economy, the economic basis of the area. They lack the nomadic education. To be a nomadic herdsman requires knowledge of vast areas of land like the palm of one's hand, to know which of many plants are fodder for the animals or poisonous, or which are useful for making tools. It requires familiarity with the behaviour of predators and parasites. To be a leader among herdsmen also requires knowledge of history and the local equivalent of sociology. One needs to be aware of interethnic clan relationships, wherever they exist, or of other far-reaching relationships (distant origins, religious ties), to be able to appeal to them in times of war or drought. One needs to keep track of the stock partnerships of one's father and grandfather, of the animals they gave out as gifts and loans, in order to be able to claim animals in return. To acquire this knowledge, the children of nomads need to stay in touch with their families. And that should not be achieved by restricting the movement of households to a narrow radius around permanent settlements where a school can be found, as is frequently the case, leading to fast deterioration of the pastures within that reach. The solution would be to provide formal education, reading, writing, mathematics and the rest, on a mobile basis. Rather than building classrooms, a lorry should be given to a school and it should move with the nomads, the way Qur'an schools have always done, even without lorries. Thus one would be able to train the future producers of milk and meat and contribute to their integration into the national economy and society, rather than producing school leavers for an urban job market, which cannot absorb them.

In by far the largest part of the lowlands there is no basis for sedentarization[1]. Agriculture is not possible, and in many of the mushrooming rural towns and trading centres of northern Kenya people either live on handouts or trade that is split into smaller and smaller portions as more people move into this sector without an increase in the overall buying power[2]. Other people are miserable servants of small-scale traders or underpaid teachers. There is a constant demand for stimulants like beer and *miraa* (*khat, qat, qaad, catha edulis*), but even that is limited because there is little wealth generated locally to acquire these or other goods. True, some people receive a salary. But how much money does a teacher have available to drown his frustrations?

To enable people to have productive lives and to help them out of the sedentarization trap, education needs to be provided in a mobile form. What is true for education also applies to other services like health and security. Basic health care should be put on wheels.[3]

Insecurity leads to overgrazing around the urban centres and in all those areas to where people withdraw. There, due to trampling and overuse of the most palatable species, pasture regeneration is actually impaired in such a way that, come the rain, pasture production is less than proportional to precipitation. On the other hand insecurity leads to vast tracts of land becoming no-man's land between the grazing areas of the opposed groups, areas where no one dares to go. There, undergrazing also

[1] This is not meant to imply environmental determinism. No environment imposes just one specific way of life. In an arid environment such as northern Kenya, one can also work in tourism, mining, software development or many other branches, although many of these options might be taken up more easily by people other than those who populate northern Kenya currently. But, if one limits the discussion to food production, mobile forms of animal production seem to be the only option, apart from hunting and gathering (on ecological variations cf. Adano and Witsenburg 2005, 2008).

[2] There is a NGO and development discourse going on about 'income generating strategies', 'coping mechanisms', and the 'ingenuity' and 'agency' of impoverished, sedentarized nomads. Insofar as such traits do not necessarily increase overall production and buying power, accounts of these developments may be too optimistic. A limited market comes into being with the influx of money from outside for handicrafts produced for tourists. This is accompanied by the usual process of self-folklorization. Miniature weapons and 'traditional' bead ornaments are produced for tourists' tastes and needs.

[3] None of these recommendations are new. Similar things have been said before by numerous authors, including myself. I am just reiterating earlier recommendations, which have found their way even into a publication by the Kenyan Government (Ministry of Livestock Development), the *Kenya Range Management Handbook* (Shaabani et al. 1991, 1992a–d).

leads to the destruction of pastures, to bush encroachment and the invasion by parasites of all sizes, from lions to tsetse flies, which find cover in the bush. For a sustainable form of nomadic land-use in these spots threatened by degradation, more animals need to move away for longer periods, mobility needs to be brought back to former levels, and for that purpose security needs to be brought to where the vegetation is. Away from the concentrations of population, where pasture regeneration solely depends on precipitation due to the erratic spatial distribution of rainfall, the only rational system of utilization is characterized by opportunistic (rather than periodic and following regular circuits), fast, and wide-ranging mobility. The policies towards pastoralists have to take these natural givens into account.

Even if the government can be persuaded to provide services on a mobile basis, that would not be sufficient. The nomadic population must group itself into units, which are sufficiently large to absorb these inputs. Not every hamlet of a dozen households can be accompanied by a teacher and a health worker, but it should be possible to provide units of about two hundred dwellings (mat-covered tents) with these services and also with some administration police and enough registered guns with counted (and to be accounted for!) ammunition to deter raiders or deal with them appropriately. The self-administration of such units should work hand in hand with the public administration and other such units to coordinate grazing management and maintain mobility and security. With this degree of co-ordination, it should also be possible to limit grazing pressure around wells and bore-holes. Without it, water development often has adverse effects: the vegetation around new bore-holes tends to deteriorate fast. Therefore, regulation is required, but the system needs to be kept flexible, so as not to interfere with basic liberties. It must be possible for individuals, households, or lineages to leave pastoral units and join others.

Critics of development intervention have often stressed that the 'traditional' systems of pastoral production are those that work the best. This may be true when one compares them with systems which suffer the effects of misconceived development intervention,[4] but it does not mean that the 'traditional' systems are the best of all possible systems. Mobile animal husbandry has become associated with backwardness. Often, this backwardness has been romanticized. The Maasai and the Hima (of Uganda) are among the preferred objects of photography because of their 'archaic' cultures, but now we should come to regard this image of nomadism as itself outdated. Modern technology has overcome the stage when it required a sedentary form of life. Portable computers and satellite dishes enable us to communicate with the rest of the world wherever we are. If a mobile education unit wants to have a library, it no longer needs

---

[4] In many cases I know, development intervention has had no lasting effects, neither positive nor negative, and can therefore not be regarded as dangerous. A couple of years after the money has been spent and the reports written, local people are not able to identify any way in which the project had altered their way of life. For a history of development intervention in Marsabit District see Machan 1999.

an extra lorry for the books. A substantial library can be stored in the memory of a computer or be carried along on a couple of CDs, with internet access even that is not necessary. Modern communications technology could enormously facilitate the provision of mobile services, including education, health, and security. That nomads are still lured into town by the sedentary provision of these services, is an anachronism. To become 'civilized' used to involve to become settled. The place of modernity was the city. All this no longer needs to be the case. Not nomadism but sedentism has become outdated.

Among Kirill Istomin's helpful comments on an earlier version of this chapter were some critical remarks about mobile services. He points to the history of failures of mobile education and mobile health care in the Russian north among reindeer nomads of the tundra since the 1920s. The basic reason for these failures was the dispersal of the nomads. In response to my suggestion that the nomads should combine into units of circa 200 households, he notes that the combined reindeer herd of such a large nomadic hamlet would not be manageable. Reindeer nomads would form groups of up to twenty households and then there would be too few children to justify the employment of a teacher who moves with them, even if one combines them all in one class. These points are perfectly valid, and I am far from propagating universal solutions.

But among the Gabra of the northern Chalbi region of Kenya, 200 households were combined into one nomadic unit in colonial times for security reasons. A camel patrol was attached to each such unit. Larger units, of course, have to move more often than smaller ones, because the pasture in their vicinity is exhausted faster, but that is what nomadism is all about. An even distribution of smaller camps over a large area would not increase the pasture but just use it up more slowly and then finish it in the entire area. It would be affordable to attach a teacher to such a unit and also a health worker, who would not replace a doctor but whose efforts would considerably reduce the frequency of people having to see a doctor. Even if nomadic education under such conditions would not lead to the same school certificates in the same time as boarding schools would do, it would still provide formal education to children who would otherwise receive no formal education at all. They could also help postpone the transfer of those who want to pursue a fuller educational career to a more mature age, after acquiring some pastoral skills and some pastoral knowledge, so that the educated elites are not alienated from pastoral life and can later become useful for their own communities. Istomin also points to alternative solutions like taking teachers from ordinary schools to nomadic children during weekends or holidays or taking nomadic children to ordinary schools for such periods. Maybe one should not discuss these things as alternatives. Why not combine one with the other? The aim should be to generate as much useful knowledge as possible in combining formal education with the experience of growing up as a pastoralist.

In the colonial literature and among development agents, pastoral 149 nomads have a reputation of being stubborn and conservative. The reason for this is that the innovations brought to them were not in their interest. Other innovations are. Without any outside intervention, mobile phones have spread even faster among pastoral nomads than the necessary networks. People had phones already, before the areas they lived in were adequately covered by networks and they climbed trees or hilltops to make calls. So far, they use phones only to contact relatives in town. But with adequate coverage, mobile phones and other small, portable technologies like GPS can also revolutionize herding techniques.[5] They might help to locate a lost herd rather than dying of thirst looking for it. Nomads have a good sense for things which are portable and useful. I have no worries that nomadic children will take to portable information technologies in no time and that this will open new chances for combining a pastoral way of life with knowledge about the modern world.

Education and health care thus may no longer need to be disincentives to pastoral mobility. The same applies to security considerations. The political responses to raiding and ethnic clashes often involve the carving-up of the pastoral lands and granting rights to a given territory to a given tribe or 'community', if one prefers the politically correct label. If the administration and the provision of security were actually brought to the people, on a mobile basis to a mobile population, it might no longer be necessary to enforce district boundaries which date from the colonial period. There would be more efficient means of guaranteeing peace than patrolling a boundary, which cannot be effectively patrolled anyhow. Movements could be organized to take into account the needs of pasture, water, and minerals of the livestock of one's own and the neighbouring pastoral units rather than obeying lines on the map, which were drawn generations ago and do not take into account any of these things.

The current political trend in Kenya, however, goes precisely the opposite way. Neither organization skills nor technical innovations are used to enhance mobility and the optimal utilization of marginal lands. On the contrary, everything is done to subdivide territories, which are already too small, into even smaller units. Prior to the last national elections in December 2007, numerous new districts were created and constituencies were about to be subdivided, a process stopped by a parliamentary committee. Subdivision at all levels was just what the political 'elites' had asked for and the government, believing this to buy loyalty and support, was only too willing to comply. Kenyans at present expect this process to continue, some critically, others full of hope for their particularistic interests.

Kenyan Members of Parliament receive some of the highest salaries in the world. In addition, the newly created 'Constituency Development Funds' are at their disposition. The constituency (and the MP who won it) have become the new channel of funds, replacing the District and the earlier policy, introduced by Moi in 1983, of the District Focus for

[5] For an interesting discussion of technology in connection with the potential of pastoral production cf. Dwyer and Istomin 2009.

Rural Development (DFRD). In this older model, a District Development Committee, chaired by the District Commissioner (DC)[6], would submit plans to the central government, which would ultimately make the decisions (Mutie 2009). The new policy, moving away from the administrative units of the executive to those of the legislation (although the two often coincided in their geographical extension), has given the MPs the opportunity to ingratiate themselves not only with their constituencies in the wider sense (the electoral district and its population) but to specifically favour their constituencies in the narrow sense of whoever voted for them, and to discriminate against clans, lineages, or individuals who supported their rivals. Social disruption thus follows the creation of smaller and smaller 'Bantustans', to take up the apartheid metaphor once more, in which specific groups and networks claim privileges at the expense of others.

In many cases the pastoralists quite clearly perceive that it would be in their own interest to enhance the productivity or their herds by an open boundary policy and by reaching agreements with their neighbours, granting each other access to pasture and water. But few people among the more vocal and better connected 'urban'[7] population seem to care about the wellbeing of the productive elements among their district population and the economic viability of their districts. The whole logic of Kenyan politics seems to be geared toward taking and distributing (keeping what one can to oneself), not producing and selling. Mobile livestock production may be the best way to use the arid half of Kenya's surface, and with all sorts of technologies becoming smaller and easier to carry, it no longer has to imply 'backwardness' and lack of modern education. It is also what pastoralists are best at. But their representatives do not seem to care.

At the time of up-dating and expanding this text (July 2009) for the present publication, it appeared to me that the problems discussed in the preceding paragraphs (productivity of pastoral production for the national economy, education and political integration) are almost luxury problems. The people around me (at Korr, Laisamis District, Rendilleland) have much more basic problems. They face the almost total loss of their livestock, and even their camels are affected by the drought at a time when the dry season can be expected to drag on for another five months. There is also no guarantee that the next rainy season will actually occur nor that rains, if they come, will actually fall here. Animals are dying, while people suffer malnutrition if not starvation. There have been a number of really bad years in my experience among the Rendille, 1984 being one of them, and this is one more such year with the potential of becoming one of the worst.

This place, to where Rendille have been lured by mission hand-outs and stationary schools, and where they have become sedentary, neglecting even to train loading camels for the eventuality that they

---

[6] DCs are appointed by the central government and come from all over the country. The ethnic group of the incumbent president has always been over-represented among them.

[7] 'Urban' is here used in a relative sense. The 'towns' of northern Kenya are often the size of villages elsewhere.

will have to move at one point, may turn into their destiny of an all too final sort. Even in normal years, when the camels, ewes, and goats have plenty of milk, the milk is not where the children are. With the exception of brief periods (if and when a rainy season produces enough pasture in the vicinity and for the sacrificial *sooriyo* ceremonies four times per year), the herds are kept in far away satellite camps with some young adults. These camps move hundreds of kilometres away because, after all, ecological givens are givens. This pattern has existed for a long time (cf. Shaabani et al. 1991). The old, the married, and the children stay here, living on money from the occasional sale of livestock and from remittances from relatives who work as watchmen, mainly in Nairobi. Proceeds from livestock sales decline. The livestock is no longer fit for sale, or dies in far away places. The diet of 'urban' Rendille used to consist mostly of carbohydrates, maize meal for *posho*[8], and sugar (for tea) for most of the year (Roth et al. 2005 p. 175), and now there is not enough even of that for many, because the money to buy these items is missing. The gap is only partly filled by famine relief.

In the new system, in which herds are kept in satellite camps for much longer periods than they were before, one can again observe that one form of mobility compensates for another.[9] The mobility of the animals remains roughly the same as it was before, because, after all, the animals need to go where there is pasture and water. But the movement of the settlements with the mat-covered houses is greatly reduced. At the same time, movements between satellite camps and settlements become more frequent (because the separation lasts for longer periods) and more demanding (because the satellite camps keep moving farther away). The most lamentable aspect of this separation, however, is that the milk tends to be in one place and the children in another.

Donkeys have always been used to accompany the herds of smallstock and to carry provisions and water for the herders. For transporting houses and fetching water for the hamlets, camels were used. These loading camels have now been given up. Donkeys and people have taken over

---

[8] *Posho* is stiff maize porridge. A variant served in Italian restaurants is known as *polenta* – the common Kenyan term is *ugali*.

[9] On forms of pastoral movement: with the possible exception of special market niches for 'organic' foods, livestock production generally requires a certain total amount of mobility which may, however, consist in several forms of mobilization that have the potential to replace one another. If the atoms that make up the piece of beef we are eating or the milk we are drinking had their origins written on them, we would see that they stem from several locales within a larger region or even from all around the globe. This is true for all kinds of livestock production, including nomadic forms, agro-pastoralism, ranching, dairy farming, and even intensive stable-feeding and zero-grazing forms of cattle keeping in peri-urban or urban settings. All forms of livestock production require a wide geographical range and a lot of mobility. There is no such thing as sedentary livestock production. In the 'oily' kind of agriculture of the developed world, huge amounts of fossil hydro-carbons are used for cattle production. Primarily, it is burned in the combustion engines of ships and motor vehicles. Fodder crops are shipped across oceans. Beef may come from an animal that was born in one place, raised in another, fattened in a third and slaughtered in a fourth that is hundreds of kilometres away by lorry. Its meat then ends up 'fresh' in a wide range of supermarkets, or it may be frozen. If the latter is the case, and there is overproduction, it may be dumped again in Africa, from where some of its atoms (in the form of exported fodder plants) originally stem.

the transport tasks. When a hamlet has to move for a couple of hundred metres (and that is what the range of movement has been reduced to) people put their belongings onto donkeys or use their own muscles to drag them along. The donkeys now are collapsing and many have died. To transport water to the hamlets, women take over. But the system of satellite camps, with the smallstock moving in a wider radius around the waterholes than the hamlets with the houses (mat-covered tents) and families, can no longer be maintained as the pasture recedes, the distance to the wells increases and the donkeys die. The smallstock herds, which may survive to a larger extent, are those that have moved far west into Samburu District or to the vicinity of Isiolo. There they can still find pasture in a manageable distance from water points. And fortunately, the new system of territorialized ethnicity has not yet started to prevent Rendille herds from moving into neighbouring districts. The situation for Boran and Gabra, who at present would not dare to venture into each other's vicinity, is different.

The Rendille around me agree that in the old system, with the camel powered mobility of hamlets and families, they could all be with their herds, hundreds of kilometres away from here, where the situation is that of an ordinary drought, something they manage very well, instead of the present prospect of total loss of livestock and the choice between famine relief and starvation for the people.

The hyenas have left. Last year, when the livestock was healthy and tasty, hyenas made daring attacks to snatch sheep or goats from under the eyes of the herdsmen and were even a danger to people. Now the ubiquitous carcasses are hardly touched. Only the eyes are plucked out by crows. Not even vultures circle in the sky. Predators and scavengers are no longer interested in what is left of Rendille livestock production. One elder explains to me that hyenas are diviners. There are human diviners, the people of the lineage Chaule in Maasula, Ariaal, who are said to be able to converse with them. But there is no need to predict a disaster which has struck already. Diviners or not, hyenas certainly have enough common sense to go to where the situation is bearable. I never heard them advocate sedentism and territorial subdivision.

LOOKING ACROSS THE BORDER INTO ETHIOPIA

The ethnic groups mentioned in the beginning in many cases straddle the Kenyan/Ethiopian border. There are Gabra, Boran, and Garre on both sides. It is therefore of interest to know how ethnic issues involving the same groups are dealt with in the neighbouring country, in a different political system. This situation also invites the study of mutual influences between ethnic politics on both sides of the border.

For the southern Oromo, Haberland (1963) is still the classic reference. For the 'proper' Oromo sub-ethnic groups, the Boran, Guji, and Arsi, his work has the richness of an ethnographic and historical encyclopaedia. As 153

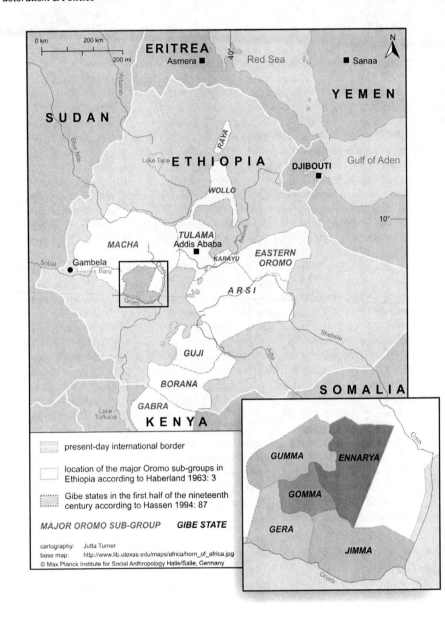

Map 4.1 Location of the major Oromo sub-groups in Ethiopia according to Haberland (1963 p. 12), and Gibe states in the first half of the nineteenth century according to Mohammed Hassen (1994 p. 87)

Haberland (1963 p. 143) points out, he had no opportunity to work among the Gabra, Sakuye or Garre, the peoples we have referred to above as Oromo-ized Somali-like, former members of the *Worr Libin* alliance or heirs of the Proto-Rendille-Somali (PRS). These neighbours of the Boran, many of whom are also Oromo in the wider, linguistic sense because they speak Oromo today, are of special interest to us here, as we are dealing with the interethnic organization of rights in pasture. Unfortunately, Haberland's information on these groups is sketchy and partly misleading. The choice of the term 'vassals' for them, which stems from European feudalism, is particularly unfortunate.

In the course of their rapid expansion in the sixteenth century, the Oromo not only came to dominate southern Ethiopia and parts of what is now Somalia and Kenya. They also penetrated into western (Wollega), northern (Wollo), and eastern (Hararghe) Ethiopia, just to mention the extremes of their expansion, so that today the Oromo-speaking area, since 1991 the regional state Oromia, extends through almost the whole width of Ethiopia. Much of this expansion took place in fertile highlands with developed agriculture, a mining industry, and fully-fledged states in the shape of little kingdoms, which were either independent political entities or entered into changing alliances with each other to subdue and pillage each other or to enforce payment of tributes. The fullest description of this process is Mohammed Hassen's work (1994), which focuses on the Gibe states, like kingdoms along the Gibe river (the northern end of the Omo catchment) by the names of (clockwise from the south) Jimma, Gera, Gomma, Gumma, and Ennarya (later Limmu-Ennarya). This region lies around 200 km south-west of the modern capital Addis Ababa.

One after the other, these states were taken over by Oromo. The strong men among the no longer pastoral Oromo thus ended up as kings of a sedentary population of mixed farmers, artisans, and slaves. The *gada* councils, formerly legislative assemblies that had age-set officials who rotated in an eight-year rhythm as their executive branch (the senior-most among them bearing the title Abba Gada), became advisory, legitimizing or purely acclamatory bodies in the service of autocratic kings.

These states, and especially the kings who did not only keep tributes for themselves but also privileges on certain types of trade and monopolies on branches of production, like musk from civet cats, for themselves, enjoyed high levels of production with surpluses for long-distance trade and high levels of consumption for the emerging upper strata at the expense of free peasants and slaves. Traces of the more egalitarian *gada* system continued to fade. Mohammed Hassen depicts court life in a Gibe state marked by a coffee and mead drinking culture with leisure and wealth consisting of gold, ivory, musk, and slaves. *Chat (catha edulis)* was dipped in honey before chewing. Bees also provided the wax for lighting the royal residence (Hassen 1994 p. 123). All this is a far cry from the pastoral life and the egalitarian generation-set based bands of warriors, which had been characteristic of the Oromo in an earlier period.   155

Boundaries between these petty states were only marked but not fortified. At the interior of these states and generally in the Oromo dominated parts of the Ethiopian highlands (from the sixteenth century to the Amharic conquest under Menelik in the late nineteenth century) interethnic relations were of different kinds:

1. There was adoption of peoples who submitted to the Oromo and joined their ranks. Intermarriage also led to assimilation without any trace on a very large scale (Merid Wolde Aregay 1971 pp. 316, 418–420). By far the largest part of the ancestry of the present-day Oromo must have been non-Oromo before the sixteenth century. Being Oromo proliferated as a successful model and as a political affiliation. The genes of Oromo conquerors may have spread rapidly, but not as fast as their language and their social organization.

2. In other cases adoption did not take the form of full individual assimilation but as incorporation at the group level and the continued ascription of an inferior status (Triulzi 1996). Separate *gada* systems were set up for *gabar* (an Amharic term the Oromo had taken over) for people of dependent status (Merid Wolde Aregay 1971 pp. 416–420). In Wollega (which is the example I have heard of – there may be more) such status differences appear to endure. When I asked poor Oromo labour migrants from Wollega in the Sudan about their tribal origin, they explained to me that they were "not good Oromo" and in some cases implied that this had played a role when they had lost claims on land and were forced to migrate.

3. Apart from the rulers and the ruled, there were guests. *Jabarti*, Muslim traders from the Christian Amhara kingdom(s) to the north and north-east were courted by the kings who competed for market outlets for local products and luxury goods from abroad.

The Oromo states often were at war with one another. Modern nationalism, which perceives Ethiopian history as a struggle between Christianity and Islam, later 'pagan' Oromo against the Christian kingdom, and yet later as a Christian 'colonial' conquest of Islamic and pagan territories, leads to an anachronistic interpretation inspired by modern classifications. Since the times of the Prophet, even before the conquest of Mecca, there have been Muslims and Christians in Ethiopia and also the languages peoples speak, and their ancestral forms have a long history. But rulers, administrators, and soldiers mostly did not act on behalf of such wider religious or ethno-linguistic groups. History abounds with examples of Christian-Muslim alliances against other Christian rulers, Amhara-Oromo alliances (with the conversion of many Oromo to Islam, the categories 'Muslim' and Oromo had come to overlap) against other Oromo or Tigray, and of course individual soldiers or entire units of mercenaries of the same origin fighting for some political unit dominated by people of another ethnic or political affiliation.[10] One should not forget that the general who in the end conquered most Oromo lands for Menelik, Ras Gobana, was himself an Oromo, and – at the time – he was not regarded as a traitor of the Oromo cause. He was just an ordinary warlord and slaver, or maybe an extraordinary one among many lesser warlords and slavers. He has only become the proverbial traitor retrospectively, from the perspective of modern Oromo nationalism[11].

With regret, we have to leave the Ethiopian highlands and their fascinating history again, before they lead us too far astray from our present topic. The conclusions we can draw from them for the comparative analysis of the spatial organization of interethnic relations in the lowlands further south are quickly drawn and can be briefly stated:

1. In the highlands, the Oromo came to rule clearly marked territorial entities. Political stratification and a sedentary mode of production developed hand-in-hand.

2. The prevalent causal factors we can discern for this development are ecological economic ones. Ecologically, the highlands permitted sedentary use of forests (coffee plantations) and mixed farming, and so the Oromo came to superimpose themselves and then to become part of a sedentary population with very clear ideas about territorial boundaries.

3. 'Culture' and religion were of secondary importance in these processes. Elements of *gada* organization persisted, but were often transformed and were no longer the basis of politics and military. Islamization may have helped in the centralization of power and in the weakening of *gada* institutions. It may also have provided a state ideology. But in other Ethiopian princely states the same role was played by Christianity. So, neither having a monotheistic religion nor having a particular one of them seems to have been a decisive and indispensable factor. Both in the agricultural highlands and in the pastoral lowlands we find Muslims and non-Muslims. The wide differences in the forms of political organization and the management of space clearly result from ecological and economic factors and are only marginally affected by cultural origins[12] or religious affiliation.

The Oromo kingdoms in the Ethiopian highlands form a vivid contrast to the relationships the Boran maintained with their Somaloid neighbours further south. The latter were basically egalitarian, with a somewhat senior status, however, for the Boran. This state of affairs only ended with colonial conquest, by the Amhara north of the boundary and by the

---

[10] Examples can be multiplied. They comprise the ones cited by Merid Wolde Aregay 1971 pp. 512, 513, 571.

[11] See Shongolo 1996.

[12] According to Haberland, the Oromo, before their descent into the lowlands, may have been mixed farmers. The Oromo lowland pastoralists who penetrated the highlands in the sixteenth century may thus only have reverted to an earlier way of life (Haberland 1963 pp. 5, 774; Hassen 1994 p. 4).

British on the southern, Kenyan side. Water points were shared, with the Boran owning key water resources and controlling them, and pasture areas overlapped, but only seasonally and partly. There were also pastures too remote from the water points or too ligneous to be of use for cattle, and it is there that the 'people of the mats' (*Worr Dasse*), the Somaloid camel nomads, could stay without interfering with the cattle economy of the Boran. It is this ecological differentiation, the principle of the niche, the specialization on different species of ruminants with different needs, which reduced competition between the different groups of pastoralists and thereby was conducive to the reduction of violence and a measure of peace. The fact that this specialization is now frequently reversed by Boran acquiring camels, Gabra taking up cattle husbandry, etc. does not augur well for peace (cf. Schlee 1989a p. 51, 2009a p. 207 on 'niche'). We may summarize the difference between Oromo kingdoms and Oromo/ non-Oromo pastoral systems by characterizing the latter as much less hierarchical and as not based on the direct control of territories in the sense of bounded areas of the surface of the earth. Territoriality or possessiveness in a wider sense have been attached to routes, holy sites (Schlee 1990b, 1992d), and water points.

The histories of Ethiopia and Kenya differ significantly. Here we can only enumerate a few salient points. Ethiopia is the only African power ever to have beaten a regular European invasion army (– the Italians at Adwa in 1896; conversely Italy is the only European nation ever to lose a conventional war against an African country). For a long time, Ethiopia was not colonized. (Its critics say it even became a colonial power of its own.) This is how, not only for the Rastafarians but for the Pan-African and African liberation movement in general, it became the symbol for African independence. The Ethiopian colonial experience with Italy, in the end under Mussolini, was late, short, and brutal (1936–1943). In contrast, Kenya was first a 'protectorate' for many decades, then a colony, and in its more attractive parts a settler colony. One would therefore expect the exposure to European models of statehood to be much stronger in the Kenyan case.

Ethiopia underwent a socialist revolution in 1974, with the subsequent elimination fights between the revolutionaries, purges and all. In 1991, the socialist dictator Mengistu was ousted.[13] The new regime, the core of which was composed of former separatists from the TPLF (Tigrayan People's Liberation Front) under the new name EPRDF (Ethiopian People's Revolutionary Democratic Front) adopted a policy of ethnic federalism. In Kenya, by contrast, transitions were mostly non-violent. The Mau Mau Uprising and its suppression by the British were brutal, but did not lead directly to the independence granted in 1963. This was the result of negotiations with more moderate elements. Since independence, as we have seen above, political discourses stressed nation-building and national unity. While Ethiopia adopted a constitution in 1994 which stressed

[13] He now enjoys the hospitality of Robert Mugabe in Zimbabwe.

ethnic federalism and ethnic self-determination including the right of secession,[14] the dominant discourse in Kenya was still one of universal citizenship as Kenyans and against 'tribalism'.

In view of all these differences, it is surprising to see how similar a role ethnicity plays now in both countries. The only difference appears to be that, in Ethiopia, ethnicity has been the official organizing principle since 1991, while in Kenya it has grown into that role unofficially but with the collusion of many people. Why is this so? One factor seems to be that ethnicity, with its variants 'indigenous rights', 'minority rights', 'culture', 'cultural defence', etc. is riding on a wave of legitimacy world-wide. Not surprisingly, after the Second World War, nationalism, *Volkstum* (peoplehood), territorial claims based on ethnicity, and the like had been discredited and raised suspicion if not aversion. Modernist discourses against traditionalism, parochialism, and tribalism added to this effect. This obviously changed since the 1970s. Ethnicity once again became a good and perfectly legitimate thing. Religion has gone through similar ups and downs. Not too long ago it was regarded as a pre-modern relic that, at the most, had a place in one's private life, but since then it has re-gained its public role and its legitimacy as a point of political identification. Crusaders and Jihadists from the radical fringes of their respective communities are presently engaged in a process of verbal escalation unheard of since the conquest of Constantinople. These ups and downs follow a global tidal rhythm, and they seem to affect remote areas like southern Ethiopia and northern Kenya in the same way, accounting for some of the similarities we find in the two places.

Of course there have also been direct influences by people and ideas moving across the border. In the 1920s, the British had the idea of grossly subdividing the peoples of the northern Kenyan lowlands into two broad categories, Oromo (or 'Galla', as they were called then) and Somali. There were and are people who do not really fit into this classification, and those who do still have many historical links to the other category. Still, the British drew a territorial boundary, the 'Galla-Somali line', which can be seen on the Map 4.1. This idea seems to have caught on in Ethiopia.

The recent work of Fekadu Adugna (2009) offers great detail about this Oromo-Somali line in Ethiopia. While the Kenyan part of the line is now a provincial boundary (between Eastern and North Eastern Provinces), its continuation in Ethiopia is the contested boundary between the Oromia and Somali Regional States.

Having written so much about how historical and ethnographic data does not fit and the constructedness of the distinction, what I find most remarkable in Fekadu's findings is the extent to which the dichotomy Oromo/Somali has been accepted by local actors in recent decades. It is an imposed dichotomy. It is a bit like (correctly) stating that the major branches of the Indo-European language family along the western shore of the European continent are the Romance branch and the Germanic branch,

---

[14] Few observers would advise any group to actually attempt it.

and that the French and the Germans are the major representatives of these two branches. So far so good. But the Ethiopian logic described by Fekadu would then lead to the conclusion that the smaller peoples or nations, the Belgians, Dutch, Danes, etc. should please make up their mind whether they want to be French or German. The Dutch certainly would object, I suspect.

Around 1990, I met Garre in the Mandera District, Kenya, who insisted that they were neither Somali nor Oromo but Garre. Gerald Hanley, in his popular account of his adventures during the Second World War, *Warriors and Strangers* (1971), describes the pride the Somali take in being Somali, and then, in contrastive terms, discusses his servant and guide, a Garre, as a person from a world in which more than anything else it mattered to be Garre, with the implication that this was an alternative identification and that the person in question would not understand why the people in Somalia, or anyone else for that matter, would care about being Somali, as it was clearly more desirable to be Garre. What has become of all the people of northern Kenya and southern Ethiopia who claimed to be neither Oromo nor Somali? Today, they identify as one of the two, and even if they switch back and forth they only claim to be one of the two at a time. *Tertium non datur.*

When I spent many hours in the Office of the President in Nairobi to get a research permit in 1978 for my research on interethnic clan relationships in northern Kenya, which later resulted in the book *Identities on the Move* (Schlee 1989a), the main difficulty was that I also wanted to visit the Somali districts. Memories of the secessionist war (the *shifta* emergency) were still fresh and I was told that since then no one had ever received clearance for research on Somali in Kenya. In order to dispel the fear that my research might be too close to politics, I stressed its historical aspects and even said that I was more interested in the sixteenth century than in the twentieth, as the sixteenth century, which witnessed the Oromo expansion, was thought to be a formative period for the present ethnic groups. Little did I know that what I said in order to appear harmless, namely my interest in ethnogenesis and 'origins', was soon to turn into one of the most hotly contested political issues. Even the sixteenth century has turned into a minefield. People are abused, threatened, misquoted, accused, and slandered on the internet if they discover something which others do not want to hear.

As ethnicity in the region then was much less politicized than it is now, it may have been easier to use the critical historical methods to get behind the individual discourses and to get a clearer idea of what really might have happened. Apart from analysing discourses as interest-guided, as Fekadu does, I attempted to reconstruct history from my perspective, as a result of a comparison and critical analysis of a great variety of sources. My reconstruction did not result in a binary opposition between Oromo, on the one hand, and Somali, on the other. I found important cultural elements of the modern pastoral groups to go back to a Proto-Rendille-Somali complex.

Apart from the Rendille, Sakuye, and others, I included the Gabra and Garri discussed by Fekadu in this perspective. My conclusions were that those among the ancestral populations, who were bearers of this PRS culture, not Somali, but speakers of Somali-like (Somaloid) languages like Rendille. For being Somali, they lacked the core feature of being Muslims. I thought and continue to think that at an early time and before their split into the modern ethnic groups, they may have borrowed elements of Arabo-Islamic culture, but that the bulk of their camel-oriented beliefs and rituals had (and in many cases still have) nothing to do with Islam. They were not Oromo either. Many of them adopted the Oromo language (or more precisely the Boran dialect of it) only in recent centuries.

Still, although among many other influences, I identify *three* major interpenetrating cultural strata in the region: Oromo, PRS, and 'modern Somality'. By the time my book was out, however, it was read from the perspective of a binary division. Oromo accused me of Somali sympathies (by misreading PRS as 'Somali' and concluding that I attributed Somali origins to so many people), and Somali, including my own wife, criticized me for not sharing their belief that the Rendille are just a lost group of Somali and at one point of time were all Muslims. The world had suddenly changed in a way so that everyone had to be either Oromo or Somali with the effect that more complex findings were not even understood.

I have written about the colonial 'Galla[15]-Somali line' that divided northern Kenya since the 1920s as a colonial imposition, which did not really fit the local givens and was therefore circumvented in many ways. People identified as Oromo or Somali to fit the categories of the British. For themselves, they knew better. I have observed the emergence of 'Oromo' and 'Somali' as politicized identifications over several decades. If there is anyone who should not be surprised by Fekadu's findings, it is me. Still, I was fascinated and even to some extent shocked by his findings. I would not have anticipated the two categories becoming so dominant as not to leave room for anything else in such a relatively short time. The most differentiated perspective Fekadu records is that of Gabra and Garre elders claiming to be Somali by origin and Oromo by culture. Of course there is a grain of truth in this. They themselves might share more features with the 'Oromo' (in this context always meaning the Boran) than their ances-tors who might have more widely spoken Somaloid languages and were 'Somali' in some wider sort of sense. This perspective is already quite sophisticated and might be difficult to maintain in a political environment, which favours simplistic slogans. Still, differentiation and scepticism do not go far enough to question the categories 'Oromo' and 'Somali' as such. What justifies referring to ancestral populations, who lived 500 years ago, by the names of modern ethnic groups? How far can these identities be traced back, and were they the only ones in earlier periods of time? The – quite plausible or at least arguable – position that the ancestral Garre or Gabra were neither Oromo nor Somali, but something intermediate or

[15] Galla is an old name for the Oromo, which since has become politically incorrect.

different from both, apparently is no longer upheld by anyone. In 1990 in Mandera, I might have met some of the last believers in a Garre identity, which is separate from both Oromo and Somali.

Fekadu analyses the factors that favoured this process as an increasingly rigorous framing of social identities with great detail. Selective historical memories of things are responses to political incentives. These incentives are of a different kind for pastoralists and for the educated 'elites', i.e. those who speculate on finding employment in administration and politics. While the former, once the principle of granting each other access to water points and pastures is given up, are interested in reserving strategic water points and thereby access to pastures for themselves, for the 'elites' it is not so much the whereabouts but the existence of a boundary that is essential. They need separate political units to the name of their own group to acquire a new set of administrative functions earmarked for them. Like in Kenya, the process of territorial subdivision also in Ethiopia is 'elite'-driven. Pastoralists simply adjust to the new rules of the game. If it is about dividing the cake, one has to be sure to get one's share.

In the end, pastoral production will suffer, because pasture becomes too restricted to balance climatic risks and seasonal fluctuations. Less meat and less milk will mean less food and less money and increased competition with loss of human life and less potential for the survivors to develop their full humanity. It is as simple as that. No grass. No food. No life. But lots and lots of politics.

In order not to end on this triad of immobility, malnutrition and exclusion, we may look for some signs of hope. There are positive developments in the field of politics and policy. We have mentioned the new Kenyan Constitution which will – if the state is not hijacked by people who ignore the constitution or just pay lip service to it – strengthen the rights of citizens also in the northern half of the country and enable them to participate more fully in democratic decision processes.

Democracy sounds like a very general issue which has little to do with the preservation of pastoral mobility. But it has to do a lot with it. The continuing tendency to territorial subdivision and obstruction of nomadic movements results, as we have seen, from the fact that policies are made by urban 'elites' who have no stake in livestock production. Broadening the democratic base and including the actual producers in decision processes which affect the ways production is organized might make a difference. We have, however, seen that a key issue for pastoralists, their collective land-rights, has not been addressed directly by the Constitution. All we can hope for is that the better inclusion of pastoralists in the political system will open the way to addressing this problem in the future.

Inclusion of pastoralists in the political system in this context does not mean their right to vote. They have that already. What we mean is their full participation in shaping politics and this requires modern education.

At this point we can come back to another document, namely the

References

Adano Wario Roba and Karen Witsenburg, 2005. 'Once Nomads Settle: Assessing Process, Motives and Welfare Changes on Marsabit Mountain', in Elliot Fratkin and Eric A. Roth (eds) *As Pastoralists Settle: Social, Health, and Economic Consequence in Marsabit District Kenya*. New York: Kluwer Academic/Plenum Publishers, 105–36.

— 2008. *Pastoral Sedentarisation, Natural Resource Management, and Livelihood Diversification in Marsabit District, Northern Kenya*. Lampeter: Mellen Press.

African Union (AU), 2010. *A Policy Framework for Pastoralism in Africa. Securing, Protecting and Improving the Lives, Livelihoods and Rights of Pastoralist Communities*. Addis Ababa: Department of Rural Economy and Agriculture. www.africa-union.org.

Ajmone-Marsan, Paolo, José Fernando Garcia, Johannes A. Lenstra, and the Globaldiv Consortium, 2010. 'On the Origin of Cattle: How Aurochs Became Cattle and Colonized the World'. *Evolutionary Anthropology* 19: 148–157.

Almagor, Uri, 1978. *Pastoral Partners*. Manchester: Manchester University Press.

Appadurai, Arjun, 1991. 'Global Ethnoscapes: Notes and Queries for a Transnational Anthropology', in Richard G. Fox (ed.), *Recapturing Anthropology: Working in the Present*. Santa Fe: School of American Research Press, 191–210.

Assmann, Aleida, 1999. *Erinnerungsräume: Formen und Wandlungen des kulturellen Gedächtnisses*. München: Beck.

Assmann, Jan, 2000 [1992]. *Das kulturelle Gedächtnis: Schrift, Erinnerung und politische Identität in frühen Hochkulturen*. München: Beck.

Bassi, Marco, 2005. *Decisions in the Shade: Political and Judicial Processes among the Oromo-Borana*. Trenton, NJ: Red Sea Press.

Baxter, Paul T.W., 1979. 'Boran Age-sets and Warfare', *Senri Ethnological Studies* 3: 69–95.

Baxter, Paul T.W. and Uri Almagor (eds), 1978. *Age, Generation and Time: Some Features of East African Age Organisations*. London: Hurst.

Baxter Paul T.W., Jan Hultin and Alessandro Triulzi, 1996. 'Introduction', in Paul T. W. Baxter, Jan Hultin and Alessandro Triulzi (eds), *Being and Becoming Oromo: Historical and Anthropological Enquiries*. Uppsala: Nordiska Afrikainstitutet, 7–25.

Behnke, Roy H., Jan Scoones, Carol Kerven, 1993. *Range Ecology at Disequilibrium: New Models of Natural Variability and Pastoral Adaptation in African Savannas*. London: Overseas Development Institute.

Berliner, David, 2004. 'The Abuses of Memory'. Presentation at the European Association of Social Anthropologists 8th Biennial Conference, September 8–12. (Book of Abstracts No. 46)

Bonte, Pierre, 1981. 'Les éleveurs d'Afrique de l'Est sont-ils égalitaires?', *Production pastorale et société* 9: 23–37.

Broch-Due, Vigdis, 1999. 'Remembered Cattle, Forgotten People: The Morality of Exchange and the Exclusion of the Turkana Poor', in David M. Anderson and Vigdis Broch-Due (eds), *The Poor Are Not Us: Poverty and Pastoralism in Eastern Africa*. Oxford: James Currey, 50–88.

Carr-Hill, Roy, 2006. 'Educational Services and Nomadic Groups in Djibouti, Eritrea, Ethiopia, Kenya, Tanzania and Uganda', in Caroline Dyer (ed), *The Education of Nomadic Peoples: Current Issues, Future Prospects*. Oxford, New York: Berghahn Books, 35–52.

Chenevix Trench, Charles, 1993. *Men Who Ruled Kenya: The Kenya Administration, 1892-1963*. London: Tauris/ Radcliffe Press.

Committee of Experts, 2010. 'The proposed Constitution of Kenya published by the Attorney-General in Accordance with Section 34 of the Constitution of Kenya Review Act, 2008 (No.9 of 2008)'. Nairobi Kenya: Attorney-General. Final version as promulgated can be seen at www.kenya-information-guide.com/kenya-constitution.html.

de Jode, Helen (ed), 2009. *Modern and Mobile. The Future of Livestock Production in Africa's Drylands*. London: International Institute for Environment and Development, and SOS Sahel International UK,

Donahoe, Brian, John Eidson, Dereje Feyissa, Veronika Fuest, Markus V. Hoehne, Boris Nieswand, Günther Schlee and Olaf Zenker, 2009. 'The Formation and Mobilization of Collective Identities in Situations of Conflict and Integration'. *Max Planck Institute for Social Anthropology Working Paper No. 116*. Halle/Saale: Max Planck Institute for Social Anthropology.

Dwyer, Marc J. and Kirill Istomin, 2009. 'Technological Carrying Capacity Renders Ecological Carrying Capacity a Redundant Concept in Pastoralist Systems? A study of "overgrazing" among Komi and Nenets reindeer herders', in David Knaute and Sacha Kajan (eds), *Sustainability in Karamoja. Rethinking the terms of global sustainability in a crisis region of Africa*. Köln: Rüdiger Köppe, 271–87.

Dyer, Caroline (ed), 2006. *The Education of Nomadic Peoples: Current Issues, Future Prospects*. Oxford, New York: Berghahn Books.

Ensminger, Jean, 1992. *Making a Market: The Institutional Transformation of an African Society*. Cambridge: Cambridge University Press.

Farah, Mohamed I. 1993. 'From Ethnic Response to Clan Identity. A Study of State Penetration among the Somali Nomadic Pastoral Society of Northeastern Kenya'. *Studia Sociologica Upsaliensia* 35. Uppsala: Acta Universitatis Upsaliensis.

Fekadu Adugna, 2009. 'Negotiating Identity: Politics of Identification among the Borana, Gabra and Garri around the Oromo-Somali Boundary in Southern Ethiopia'. Ph.D. thesis, Halle (Saale): Martin-Luther-Universität Halle-Wittenberg.

Galaty, John, 2009. 'Walling In or Walling Out? Land Demarcation and the Genesis of Pastoral Conflict in East Africa', Paper presented at the Workshop of the Max Planck Institute for Social Anthropology in cooperation with and held at the Egerton University, Kenya, on *Ethnicization of Politics, Governance in the Borderlands, and the State in the Horn of Africa*, 7–9 July 2009.

Government of Kenya (GOK), 1981. *Kenya Population Census 1979: Vol. I*. Nairobi: Central Bureau of Statistics, Ministry of Economic Planning and Development.

Grum, Anders, 1976. 'When the Fly once Owned the Camels and 81 Other Stories of the Rendille'. (Hectographed collection of Rendille stories).

Haberland, Eike, 1963. *Galla Süd-Äthiopiens*. Stuttgart: Kohlhammer.
Hahn, E., 1891. 'Waren die Menschen der Urzeit zwischen der Jägerstufe und der Stufe des Ackerbaus Nomaden?' *Das Ausland* 64: 481–87.
— 1892. 'Die Wirtschaftsformen der Erde'. *Petermanns Geographische Mitteilungen* 38: 8–12.
— 1896. *Die Haustiere und ihre Beziehung zur Wirtschaft des Menschen: Eine Geographische Studie*. Leipzig: Duncker & Humblot.
— 1911. 'Wirtschaftliches zur Prähistorie'. *Zeitschrift für Ethnologie* 43: 821–40.
— 1913. 'Die Hirtenvölker in Asien und Afrika'. *Geographische Zeitschrift* 19: 305–19; 369–82.
— 1925. 'Dreistufentheorie', in M. Ebert (ed), *Reallexikon der Vorgeschichte*, Vol. 2. Berlin, 460–62.
— 1927. 'Nomaden', in M. Ebert (ed), *Reallexikon der Vorgeschichte*, Vol. 8. Berlin, 547–48.
Halbwachs, Maurice, 1997 [1950]. *La mémoire collective*. Paris: Éditions Albin Michel.
Hanley, Gerald, 1971. *Warriors and Strangers*. London: Hamish Hamilton.
Hardin, Garrett, 1968. 'The Tragedy of the Commons', *Science* 162(3859): 1243–48.
Hassen, Mohammed, 1994. *The Oromo of Ethiopia. A History 1570–1860*. Trenton, NJ: Red Sea Press.
Homewood, Katherine, 2008. *Ecology of African Pastoralist Societies*. Oxford: James Currey.
Hussein A. Mahmoud, 2003. 'The Dynamics of Cattle Trading in Northern Kenya and Southern Ethiopia: the Role of Trust and Social Relations in Market Networks'. Ph.D. thesis, University of Kentucky.
Iliffe, John, 1987. *The African Poor: a History*. Cambridge: Cambridge University Press.
Kassam, Aneesa, 2006. 'The People of the Five "Drums": Gabra Ethnohistorical Origins', *Ethnohistory* 53 (1): 173–93.
Kenya Human Rights Commission, 2008. *Foreigners at Home. The Dilemma of Citizenship in Northern Kenya*. Pamoja Tutetee Haki.
Kenya National Archives (KNA)
— *Gurreh District Annual Reports*. Nairobi: Kenya National Archives.
— *Isiolo District Annual Reports*. Nairobi: Kenya National Archives.
— *Mandera District Annual Reports*. Nairobi: Kenya National Archives.
— *Moyale Station Reports*. Nairobi: Kenya National Archives.
— *Moyale District Annual Reports*. Nairobi: Kenya National Archives.
— *Wajir District Annual Reports*. Nairobi: Kenya National Archives.
— *Wajir Political Record Books*. Nairobi: Kenya National Archives.
Khazanov, Anatoly and Günther Schlee (eds), forthcoming 2012. *Who Owns the Stock? Collective and Multiple Property Rights in Animals*. Oxford: Berghahn Books.
Kioli, Felix Ngunzo, 2009. 'Ethnicization of Political Participation in Kenya: from independence to present'. Paper presented at the Workshop of the Max Planck Institute for Social Anthropology in cooperation with and held at the Egerton University, Kenya, on the *Ethnicization of Politics, Governance in the Borderlands, and the State in the Horn of Africa*, July 7–9, 2009.
Krätli, Saverio, 2006. 'Cultural Roots of Poverty? Education and Pastoral Livelihood in Turkana and Karamoja', in Caroline Dyer (ed), *The Education of Nomadic Peoples: Current Issues, Future Prospects*. Oxford, New York: Berghahn Books, 120–40.
Krätli, Saverio and Caroline Dyer, 2009. *Mobile Pastoralists and Education: Strategic Options*. London: International Institute for Environment and Development.

Leder, Stefan and Bernhard Streck (eds), 2005. *Shifts and Drifts in Nomad-Sedentary Relations*. Wiesbaden: Reichert.

Legesse, Asmarom, 1973. *Gada: Three Approaches to the Study of African Society*. New York: Free Press.

Lentz, Carola, 2006. *Ethnicity and the Making of History in Northern Ghana*. Edinburgh: Edinburgh University Press.

Leus, Ton with Cynthia Salvadori, 2006. *Aadaa Boraanaa: a Dictionary of Borana Culture*. Addis Ababa: Shama Books.

Machan, Steve N., 1999. 'A History of Intervention in Livestock Management in Marsabit District'. A Report for the Max Planck Institute for Social Anthropology Halle/Saale, 2000. Halle (Saale): Max Planck Institute for Social Anthropology.

Mathilda's Anthropology Blog, 2009. 'The Case for and against Cattle Domestication and Sorghum Cultivation at Nabta Playa'. http://mathildasanthropologyblog.wordpress.com/2009/01/21/the-case-for-and-against-cattle-domestication-and-sorghum-cultivation-at-nabta-playa, downloaded July 2011.

Merid Wolde Aregay, 1971. 'Southern Ethiopia and the Christian Kingdom 1508–1708, with Special Reference to the Galla Migrations and their Consequences'. Unpublished Ph.D. thesis. London: University of London, School of Oriental and African Studies.

Monod, Théodore, 1975. *Pastoralism in Tropical Africa*. London: Oxford University Press, for the International African Institute.

Mutie, Pius Mutuku, 2009. 'Towards a Stable Multiethnic Kenyan State: the Kikuyu and the *majimbo* factor'. Paper presented at the Workshop of the Max Planck Institute for Social Anthropology in cooperation with and held at the Egerton University, Kenya, on the *Ethnicization of Politics, Governance in the Borderlands, and the State in the Horn of Africa*, July 7–9, 2009.

Newbury, David S., 1980. 'The Clans of Rwanda: an historical hypothesis', *Africa* 50(4): 389–403.

Niezen, Ron, 2003. *The Origins of Indigenism: Human Rights and the Politics of Identity*. Berkeley: University of California Press.

Ottaway, Marina, 1993. 'Testimony Prepared for Presentation to the House Foreign Relations Subcommittee on African Affairs Hearing'. *U.S. Foreign Assistance and Policy Issues Towards Central and Eastern Africa*.

Raikes, Philip L., 1981. *Livestock Development and Policy in East Africa*. Uppsala: Scandinavian Institute of African Studies.

Robinson, Paul, 1985. 'Gabbra Nomadic Pastoralism in Nineteenth and Twentieth Century Northern Kenya: Strategies for Survival in a Marginal Environment'. Unpublished Ph.D. thesis. Ann Arbor: Northwestern University.

Roth, Eric A., Martha A. Nathan and Elliot Fratkin, 2005. 'The Effects of Pastoral Sedentarization on Children's Growth and Nutrition among Ariaal and Rendille in Northern Kenya', in Elliot Fratkin and Eric A. Roth (eds), *As Pastoralists Settle: Social, Health, and Economic Consequences of Pastoral Sedentarization in Marsabit District, Kenya*. New York: Kluwer Academic/Plenum Publishers, 173–91.

Sato, Shun, 1996. 'The Commercial Herding System among the Garri', in S. Sato (ed), *Essays in Northeast African Studies*. Senri Ethnological Studies no. 43. Osaka: National Museum of Ethnology, 275–94.

Schlee, Günther, 1979. *Das Glaubens-und Sozialsystem der Rendille. Kamelnomaden Nordkenias*. Berlin: Reimer.

— 1982a. 'Zielkonflikte und Zielvereinheitlichung zwischen Entwicklungsplanung und Wanderhirten in Ostafrika', in Fred Scholz and Jörg Janzen (eds)

Nomadismus – ein Entwicklungsproblem? Abhandlungen des Geographischen Instituts, Anthropogeographie. Berlin: Reimer, 33: 96–109.

— 1982b. 'Annahme und Ablehnung von Christentum und Islam bei den Rendille in Nordkenia' in Niels-Peter Moritzen (ed), *Ostafrikanische Völker zwischen Mission und Regierung*. Erlangen: Lehrstuhl für Missionswissenschaften, 101–30.

— 1984a. 'Nomaden und Staat. Das Beispiel Nordkenia'. *Sociologus* 34: 160–61.

— 1984b. 'Une société pastorale pluriethnique: Oromo et Somalis au Nord du Kenya'. *Production pastorale et société* 15: 21–39.

— 1985. 'Interethnic Clan-relationships among Cushitic Speaking Groups of Northern Kenya'. Habilitation Thesis, Universität Bayreuth.

— 1987. 'Somaloid History: Oral Tradition, Kulturgeschichte and Historical Linguistics in an Area of Oromo/Somaloid Interaction', in Herrmann Jungraithmayr and Walter W. Müller (eds), *Proceedings of the Fourth International Hamito-Semitic Congress, Marburg, 20–22 September 1983*. Amsterdam: John Benjamins B.V. 265–315.

— 1988a. 'L'islamisation du passé: à propos de l'effet réactif de la conversion de groupes somalis et somaloïdes à l'islam sur la représentation de l'histoire dans leurs traditions orales', in Herrmann Jungraithmayr, Wilhem J. G. Möhlig, and Josef F. Thiel (eds), *La littérature orale en Afrique comme source pour la découverte des cultures traditionnelles*. Berlin: Reimer, 269–99.

— 1988b. 'Camel Management Strategies and Attitudes towards Camels in the Horn', in Jeffrey C. Stone (ed), *The Exploitation of Animals in Africa*. Aberdeen: Aberdeen University, African Studies Group, 143–54.

— 1989a. *Identities on the Move: Clanship and Pastoralism in Northern Kenya*. Manchester and New York: Manchester University Press, St. Martin's Press.

— 1989b. 'The Orientation of Progress: Conflicting Aims and Strategies of Pastoral Nomads and Development Agents in East Africa. A Problem Survey', in E. Linnebuhr (ed), *Transition and Continuity in East Africa and Beyond. In Memoriam David Miller*, Bayreuth African Studies Series: Special Issue. Bayreuth: E. Breitinger, 397–450.

— 1989c. 'Nomadische Territorialrechte: Das Beispiel des kenianisch-äthiopischen Grenzlandes', *Die Erde* 2: 131–38.

— 1990a. 'Holy Grounds' in Paul T.W. Baxter and Richard Hogg (eds), *Property Poverty and People: Changing Rights in Property and Problems of Pastoral Development*. Manchester: Department of Social Anthropology and International Development Centre, 45–54.

— 1990b. 'Policies and Boundaries: Perceptions of Space and Control of Markets in a Mobile Livestock Economy'. *Working Paper* No. 133. Bielefeld: University of Bielefeld, Sociology of Development Research Centre.

— 1991a. 'Erfahrungen nordkenianischer Wanderhirten mit dem kolonialen und postkolonialen Staat', in Fred Scholz (ed), *Nomaden, mobile Tierhaltung: zur gegenwärtigen Lage von Nomaden und Chancen mobiler Tierhaltung*. Berlin: Das arabische Buch, 131–56.

— 1991b. 'Loanwords in Oromo and Rendille as a Mirror of Past Interethnic Relations'. *Working Paper* No. 159. Bielefeld: University of Bielefeld, Sociology of Development Research Centre.

— 1991c. 'Traditional Pastoralists – Land Use Strategies', in Salim B. Shaabani, Markus Walsh, Dennis J. Herlocker and Dierk Walther (eds), *Range Management Handbook of Kenya: Marsabit District*. Vol. II, 1. Nairobi: Republic of Kenya, Ministry of Livestock Development (MOLD) and Gesellschaft für Technische Zusammenarbeit (GTZ), 130–64.

— 1992a. 'Traditional Pastoralists – Land Use Strategies', in Salim B. Shaabani, Markus Walsh, Dennis J. Herlocker and Dierk Walther (eds), *Range*

Management Handbook of Kenya: Mandera District. Vol. II, 4. Nairobi: Republic of Kenya, Ministry of Livestock Development (MOLD) and Gesellschaft für Technische Zusammenarbeit (GTZ), 122–35.

— 1992b. 'Who Are the Tana Orma? The Problem of their Identification in a Wider Oromo Framework'. *Working Paper* No. 170. Bielefeld: University of Bielefeld, Sociology of Development Research Centre.

— 1992c. 'Traditional Pastoralists – Land Use Strategies', in Salim B. Shaabani, Markus Walsh, Dennis J. Herlocker and Dierk Walther (eds), *Range Management Handbook of Kenya: Wajir District.* Vol. II, 3. Nairobi: Republic of Kenya, Ministry of Livestock Development (MOLD) and Gesellschaft für Technische Zusammenarbeit (GTZ), 140–49.

— 1992d. 'Ritual Topography and Ecological Use: The Gabbra of the Kenyan/ Ethiopian Borderlands', in David Parkin and Elisabeth Croll (eds), *Bush Base: Forest Farm.* London: Routledge, 110–28.

— 1994a. 'Ethnicity Emblems, Diacritical Features, Identity Markers: Some East African Examples', in David Brokensha (ed), *River of Blessings: Essays in Honor of Paul Baxter.* Syracuse, NY: Maxwell School of Citizenship and Public Affairs, 129–43.

— 1994b. 'Kuschitische Verwandtschaftssysteme in vergleichenden Perspektiven', in Thomas Geider and Raimund Kastenholz (eds), *Sprachen und Spracherzeugnisse in Afrika.* Köln: Rüdiger Köppe, 367–88.

— 1994c. 'Loanwords in Oromo and Rendille as a Mirror of Past Interethnic Relations', in Richard Fardon and Graham Furniss (eds), *African Languages, Development and the State.* London: Routledge, 191–212.

— 1997. 'Cross-cutting Ties and Interethnic Conflict: the Example of Gabbra, Oromo and Rendille', in Katsuyoshi Fukui, Eisei Kurimoto and Masayoshi Shigeta (eds), *Ethiopia in Broader Perspective. Papers of the 13th International Conference of Ethiopian Studies.* Vol. 2, Kyoto: Shokado Book Sellers, 577–96.

— 1998a. 'Some Effects on[of] a District Boundary in Kenya', in Mario I. Aguilar (ed), *The Politics of Age and Gerontocracy in Africa: Ethnographies of the Past and Memories of the Present.* Trenton, NJ: Africa World Press, 225–56.

— 1998b. 'Gada Systems on the Meta-Ethnic Level: Gabbra/Boran/Garre Interactions in the Kenyan/Ethiopian Borderland', in Eisei Kurimoto and Simon Simonse (eds), *Conflict, Age and Power in North East Africa.* Oxford: James Currey, 121–46.

— 1999. 'Nomades et l'Etat au nord du Kenya', in A. Bourgeot (ed), *Horizons nomades en Afrique sahélienne.* Paris: Karthala, 219–39.

— 2003. 'Redrawing the Map of the Horn: The Politics of Difference'. *Africa – Journal of the International African Institute,* 73(3): 343–68.

— 2005. 'Forms of Pastoralism', in Stefan Leder and Bernhard Streck (eds), *Shifts and Drifts in Nomad-Sedentary Relations.* Wiesbaden: Reichert, 17–54.

— 2006. 'The Somali Peace Process and the Search for a Legal Order', in Hans-Jörg Albrecht, Jan-Michael Simon, Hassan Rezaei, Holger-C. Rohne and Ernesto Kiza (eds), *Conflicts and Conflict Resolution in Middle Eastern Societies – Between Tradition and Modernity.* Berlin: Duncker and Humblot, 117–67.

— 2007. 'Brothers of the Boran once Again: On the Fading Popularity of Certain Somali Identities in Northern Kenya'. *Journal of Eastern African Studies* 1(3): 417–35.

— 2008a. *How Enemies Are Made: Towards a Theory of Ethnic and Religious Conflict.* Oxford: Berghahn.

— 2008b. 'The "Five Drums", Proto-Rendille-Somali, and Oromo Nationalism: A Response to Aneesa Kassam'. *Ethnohistory* 55(2): 321–30.

— 2009a. 'Changing Alliances among the Boran, Garre and Gabra in Northern Kenya and Southern Ethiopia', in Günther Schlee and Elizabeth E. Watson (eds), *Changing Identifications and Alliances in North-East Africa*, Vol. I: *Ethiopia and Kenya*. Oxford: Berghahn, 203–23.

— 2009b. 'Descent and Descent Ideologies: the Blue Nile Area (Sudan) and Northern Kenya Compared'. In Günther Schlee and Elizabeth E. Watson (eds), *Changing Identifications and Alliances in North-East Africa Volume II: Sudan, Uganda and the Ethiopia-Sudan Borderlands*. Oxford, New York: Berghahn Books, 117–35.

— 2010. 'Territorializing Ethnicity: The Imposition of a Model of Statehood on Pastoralists in Northern Kenya and Southern Ethiopia'. *Max Planck Institute for Social Anthropology Working Papers* No. 121. Halle/Saale: Max Planck Institute for Social Anthropology. (A version of this article has been released in 2011 in *Ethnic and Racial Studies* http://www.tandfonline.com/action/showAxaArticles?journalCode=rers20).

Schlee, Günther and Karaba Sahado, 2002. *Rendille Proverbs in their Social and Legal Context*. Köln: Rüdiger Köppe.

Schlee, Günther and Abdullahi Shongolo, 1995. 'Local War and its Impact on Ethnic and Religious Identification in Southern Ethiopia', *GeoJournal* 36: 7–17.

— 1996. 'Oromo Nationalist Poetry: Jarso Waaqo Qooto's Tape Recording about Political Events in Southern Oromia, 1991', in Ioan M. Lewis and Richard J. Hayward (eds), *Voice and Power: the Culture of Language in North-East Africa: Essays in Honour of B.W. Andrzejewski*. African Languages and Cultures: Supplement 3. London: School of Oriental and African Studies, 229–42.

Schlee, Günther with Abdullahi Shongolo, 2012. *Islam and Ethnicity in Northern Kenya & Southern Ethiopia*. Oxford: James Currey.

Schlee, Günther and Elizabeth E. Watson (eds), 2009. *Changing Identifications and Alliances in North-East Africa*, Vol. I and Vol. II. Oxford: Berghahn Books.

Schneider, Harold K, 1979. *Livestock and Equality in East Africa: The Economic Basis for Social Structures*. Bloomington: Indiana University Press.

Scoones, Ian, 1995. 'New Directions in Pastoral Development in Africa', in Ian Scoones (ed), *Living with Uncertainty: New Directions in Pastoral Development in Africa*. London: Intermediate Technologies Publications, 1–36.

Shaabani, Salim B., Markus Walsh, Dennis J. Herlocker and Dierk Walther (eds), 1991. *Range Management Handbook of Kenya: Marsabit District*, Vol. II, 1. Nairobi: Republic of Kenya, Ministry of Livestock Development (MOL) and Gesellschaft für Technische Zusammenarbeit (GTZ).

— 1992a. *Range Management Handbook of Kenya: Samburu District*, Vol. II, 2. Nairobi: Republic of Kenya, Ministry of Livestock Development (MOL) and Gesellschaft für Technische Zusammenarbeit (GTZ).

— 1992b. *Range Management Handbook of Kenya: Mandera District*, Vol. II, 4. Nairobi: Republic of Kenya, Ministry of Livestock Development (MOL) and Gesellschaft für Technische Zusammenarbeit (GTZ).

— 1992c. *Range Management Handbook of Kenya: Isiolo District*, Vol. II, 5. Nairobi: Republic of Kenya, Ministry of Livestock Development (MOL) and Gesellschaft für Technische Zusammenarbeit (GTZ).

— 1992d. *Range Management Handbook of Kenya: Isiolo District*. Vol. II, 5. Nairobi: Republic of Kenya, Ministry of Livestock Development (MOLD) and Gesellschaft für Technische Zusammenarbeit (GTZ).

Shongolo, Abdullahi, 1992. 'The Gumi Gaayo Assembly of the Boran: A Traditional Legislative Organ and its Relationship to the Ethiopian State and a 171

Modernizing World'. *Working Paper* No. 173. Bielefeld: University of Bielefeld, Sociology of Development Research Centre.

— 1994. 'The Gumi Gaayo Assembly of the Boran: A Traditional Legislative Organ and its Relationship to the Ethiopian State and a Modernizing World'. *Zeitschrift für Ethnologie* 119: 27–58.

— 1996. 'The Poetics of Nationalism, A Poem by Jaarso Waaqo Qoot'o', in Paul T. W. Baxter, Jan Hultin and Alessandro Triulzi (eds), *Being and Becoming Oromo: Historical and Anthropological Enquiries.* Uppsala: Nordiska Afrikainstitutet, 265–90.

— 2009. 'Interaction of Ethnicity and Factors of Land and Power in Generating the 2007 Post-election Violence in Kenya'. Paper presented in a Workshop on Ethnicization of Politics, Governance on the Borderlands, and State in the Horn of Africa. Egerton University, Kenya. 7–9 July, 2009.

Shongolo, Abdullahi and Günther Schlee, 2007. *Boran Proverbs in their Cultural Context.* Köln: Rüdiger Küppe.

Spear, Thomas T. and Waller, Richard D. (eds), 1993. *Being Maasai: Ethnicity and Identity in East Africa.* Oxford: James Currey.

Spencer, Paul, 1973. *Nomads in Alliance.* London: Oxford University Press.

Tadesse Berisso, 2009. 'Changing Alliances of Guji-Oromo and their Neighbours: State Politics and Local Factors', in Günther Schlee and Elizabeth E. Watson (eds), *Changing Identifications and Alliances in North-East Africa.* Vol. I, *Ethiopia and Kenya.* Oxford: Berghahn Books, 191–99.

Thomson, Joseph, 1968 [1885]. *Through Masailand.* London: F. Cass.

Triulzi, Alessandro, 1996. 'United and Divided. Boorana and Gabaro among the Macha Oromo in Western Ethiopia', in Paul T.W. Baxter, Jan Hultin and Alessandro Triulzi (eds), *Being and Becoming Oromo. Historical and Anthropological Enquiries.* Lawrenceville, NJ: Red Sea Press, 251–64.

United Nations, 1997. *The United Nation Convention to Combat Desertification. A New Response to an Age-Old Problem.* Earth Summit +5, Special Session of the General Assembly to Review and Appraise the Implementation of Agenda 21, New York, 23–27 June 1997. Retrieved 23 July 2011 from www.un.org/ecosocdev/geninfo/sustdev/desert.htm.

Vajda, László, 1968. *Untersuchungen zur Geschichte der Hirtenkulturen.* Wiesbaden: Harrassowitz.

Wendorf, Fred, A.E. Close and R. Schild, 1987. 'Early Domestic Cattle in the Eastern Sahara'. *Palaeoecology of Africa and the Surrounding Islands,* 18: 441–47.

Wendorf, Fred, Angela E. Close, Romuald Schild, Krystyna Wasylikowa, Rupert A. Housley, Jack R. Harlan, Halina Królik, 1992. 'Saharan Exploitation of plants 8,000 years BP'. *Nature.* 359: 721–24.

Wendorf, Fred and Romuald Schild, 1998. 'Nabta Playa and Its Role in Northeastern African Prehistory'. *Journal of Anthropological Archaeology* 17: 97–123

Wendorf, Fred, Romuald Schild, and Associates, 2001. *Holocene Settlement in the Egyptian Sahara.* Vol. 1, *The Archaeology of Nabta Playa.* New York: Kluwer Academic/Plenum Publishers.

Zitelmann, Thomas, 1990. 'Die Konstruktion kollektiver Identität im Prozeß der Flüchtlingsbewegungen am Horn von Afrika. Eine sozialanthropologische Studie am Beispiel der saba oromo (Nation der Oromo)', Ph.D. thesis, Freie Universität Berlin.

Index

173

EASTERN AFRICAN STUDIES
These titles published in the United States and Canada by Ohio University Press